A GUIDE TO FRESHWATER FISHES OF OREGON

A Guide to Freshwater Fishes of Oregon

DOUGLAS F. MARKLE

Color illustrations by Joseph R. Tomelleri

Oregon State University Press Corvallis

The John and Shirley Byrne Fund for Books on Nature and the Environment provides generous support that helps make publication of this and other Oregon State University Press books possible.

∞This paper meets the requirements of ANSI/NISO Z39.48-1992 (Permanence of Paper).

First published in 2016 by Oregon State University Press
Printed in China

Library of Congress Cataloging-in-Publication Data

Names: Markle, Douglas F., 1947- | Tomelleri, Joseph R., illustrator.
Title: A guide to freshwater fishes of Oregon / Douglas F. Markle ; color illustrations by Joseph R. Tomelleri.
Description: Corvallis : Oregon State University Press, [2016] | Includes bibliographical references and index.
Identifiers: LCCN 2016018914 | ISBN 9780870718731 (original trade pbk. : alk. paper)
Subjects: LCSH: Freshwater fishes--Oregon--Identification. | Fishes--Oregon--Identification.
Classification: LCC QL628.O7 M37 2016 | DDC 597/.7--dc23
LC record available at https://lccn.loc.gov/2016018914

Oregon State University Press
121 The Valley Library
Corvallis OR 97331-4501
541-737-3166 • fax 541-737-3170
www.osupress.oregonstate.edu

CONTENTS

ACKNOWLEDGMENTS

Carl E. Bond took me on my first freshwater field trip in Oregon to collect lamprey ammocoetes and impart his wisdom on academia, Oregon's geology, and its frustrating fishes. I am grateful to Carl and my other mentors—Ed Raney, Bob Jenkins, Jack Musick, Dan Cohen, and Bev Scott. Numerous individuals have allowed me to join their fieldwork, discussed fish taxonomy, or helped with other logistics, including Jason Adams, Brian Alfonse, Chris Allen, Pete Baki, Brian Bangs, Dick Beamish, Daryl Bingham, Carl Bond, Wayne and Patty Bowers, George Boxall, Troy Brandt, Bill Brignon, Whit Bronaugh, Abel Brumo, Brian Cannon, Jeff Dambacher, Margaret Docker, Tom Dowling, Gary Galovich, Leo Grandmontagne, Mike Gray, Stan Gregory, Mary Hanson, Phil Harris, Mike Hayes, Kendra Hoekzema, Holly Huchko, Justin Huff, Bob Hughes, Shannon Hurn, Laura Jackson, Steve Jacobs, Bob Jenkins, Jes Kettretad, Greg Kovalchuk, Michael Lance, Ron Larson, Dan Logan, Chris Lorion, Steve Mamoyac, Keith Marine, Alan Mauer, Justin Miles, Bruce Miller, Shelley Miller, Steve Naimitz, Travis Neal, Paul Olmstead, Ray Perkins, Stewart Reid, Paul Reimers, Stacy Rempel, Shannon Richardson, Chris Rombough, Tom Rumreich, Paul Scheerer, Kirk Schroeder, Brian Sidlauskas, David Simon, Brad Smith, Jerry Smith, Roger Smith, Mark Terwilliger, Bill Tinniswood, Joe Tomelleri, Dan Van Dyke, Joy Vaughan, Rick Vetter, Tim Walters, Michele Weaver, Laurie Weitkamp, Rollie White, Kelly Wildman, Randy Wildman, Doug Young, Don Zaroban, and Jeff Ziller. Several landowners—Dick Anderson, Dan Brown, and Bill and Ann Tracy—graciously allowed access to their property. Helpful reviews of early drafts were made by C. Allen, B. Bangs, S. Gregory, P. Scherrer, and R. Wildman. Specimen collections were authorized under a series of Oregon scientific taking permits, US Fish and Wildlife Service scientific taking permit TE006333-14, and OSU Institutional Animal Care and Use permits.

Financial support to initiate this work and to obtain copyright permission to use the Tomelleri images was provided by the US Fish and Wildlife Service. Generous funds from the Oregon Chapter and Western Division of the American Fisheries Society supported publication. In addition to these financial supporters, logistic support and encouragement from biologists and staff of the Oregon Department of Fish and Wildlife were also instrumental in completing the project.

Photographs of White Sturgeon, Clupeidae, American Shad, and Mississippi Silverside in the public domain by René Reyes, US Bureau of Reclamation, pages 35, 38, 96. Photographs of River Lamprey mouth and adult male © 2011 by Mike Hayes and Richard Hays, US Geological Survey, Department of the Interior, pages 30, 31. Photographs of Sacramento Perch, larval Goldfish, Blue Chub, Tui Chub, Shortnose Sucker, Western Mosquitofish, Klamath Marbled Sculpin, Klamath Lake Sculpin, Slender Sculpin, Largemouth Bass, and Yellow Perch © 2016 by Justin Huff, pages 40, 58, 59, 62, 98, 107, 109, 111, 113, 114, 123. Photographs of Entosphenus sp. and Lampetra sp. © 2016 by Greg Kovalchuk, pages 28, 30. Photograph of River Lamprey © 2016 by Laurie Weitkamp, NOAA Fisheries, page 31. Photographs of Umpqua Dace and Coho Salmon parr © 2016 by Bruce Miller, pages 48, 83. Photographs of Borax Chub and Alvord Chub © 2016 by Travis Neal, pages 56, 57. Photograph of Rainwater Killifish © 2016 by Jesse Bissette, page 97.

INTRODUCTION

In Oregon, the word "fish" means salmon or trout, members of the small but important subfamily Salmoninae. Salmon, the most iconic animals of our region, grace the headlines of our local news, and the Chinook Salmon is our state fish. These and other food fishes have long been the center of attention for Oregon fishes. Almost all the references to Columbia River fishes in the journal of Meriwether Lewis were concerned with food fishes—salmon, trout and suckers (called mullet)—and their preparation by native tribes. Oregon also boasts a large contingent of fish or fishery biologists and one of the largest and most honored state chapters of the American Fisheries Society (AFS). Not surprisingly, most of the activities of these scientists focus on salmon and trout.

There are, of course, other fishes, both native and nonnative, that live in the waters of the state. Most of these are poorly known and, in several cases, we are not even sure of the number of species we have. Some of these native fishes have become rare, and the Endangered Species Act has helped focus attention on these poorly understood species, such as the Pacific Lamprey, Oregon Chub, and Shortnose and Lost River Suckers, but there are many more needing study. In his 1994 key to Oregon freshwater fishes, Carl Bond included 109 species and subspecies, 37 nonnative and 72 native. In this work, there are 138 species and subspecies mentioned (not all are described) in 25 families, 50 nonnative and 88 native. The nonnative fishes include all species that have been found alive in state waters whether or not there is evidence of self-sustaining populations. The nonnative Highland Shiner, *Notropis micropteryx*, found as pickled bait in the Coquille River (OS 17994), is not included herein. Nonnative species have more than tripled in the last 75 years, and the inclusion of nonsustaining populations accounts for part of the increase. But there are also new, viable nonnative populations that have become established. These should be cause for concern, as nonnative fishes are frequently responsible, in part, for endangered species listings and the ensuing economic and social costs. The increase in native species and subspecies in this guide reflects new information from traditional and molecular systematics as well as a conscious effort to highlight biological diversity.

In some cases, the taxonomy of a group is not well resolved. In general, if a form is not recognized as a species by the AFS and the American Society of Ichthyologists and Herpetologists (ASIH), it is listed as a subspecies herein, though there are exceptions. There are numerous references to the need for more taxonomic work on some groups, and when that happens, the taxonomic arrangement used here may change. A conservative approach is to use the AFS-ASIH species name, recognizing, for example, that the Speckled Dace, *Rhinichthys osculus*, is most likely a complex of many species, and in the future perhaps none of the forms in Oregon will bear the name *R. osculus*.

Names of animals are important and it is worth a digression to explain how names are used in this guide. Taxonomy, the classification and naming of biological diversity, begins with naming biological species. Any zoologist who finds a species that lacks a scientific name may describe it and give it a latinized name subject to voluntary adherence to the rules and recommendations of the International Code of Zoological Nomenclature (these rules come from the International Commission on Zoological Nomenclature). The author of a scientific name is listed, with the year of publication after the name. If the genus name has changed since the original description, the author and date are in parentheses. Scientific names allow scientists to build up knowledge and communicate information about species using a common naming system. Taxonomy is also hierarchical, with names applied to larger, ever more inclusive groups of close relatives, so species are grouped into the genus category, genera into the family category, and families into the order category. However, scientific names are also one of the tools of systematics, the study of biological diversity and relationships of organisms. Systematics is an iterative discipline that seeks to create classifications reflecting evolutionary relationships. As classifications change and as new biological information accumulates, scientific names can change as well.

Common names are also important, and in North America the AFS and ASIH publish a list of standardized common names for every fish species (see Page et al. 2013). Common names can originate as folk names that arise within a community or, like scientific names, as invented appellations. A new convention, followed here, is to treat common names as proper nouns that are capitalized. In many cases there may be many different common names for a single species across its range, and selection of a single name may be difficult. Descriptive, historical, and colorful names are preferred, while offensive names are avoided. One important role of common names is to highlight biological diversity in cases where the basis for that diversity is still not clear or is not based on species differences. For example, the geographically isolated Millicoma Dace has not been given a scientific name because of uncertainty about its origins and relationships, although some recent molecular data suggest it should be recognized as a distinct species, as is done here. In salmonids, residency and anadromy create obvious differences in fish size and behaviors. It has been very useful to use common names to distinguish the anadromous Steelhead from the resident Rainbow Trout, both of which are classified as the species *Oncorhynchus mykiss*. In this work, common names will be used for species, for taxa such as Millicoma Dace that have uncertain species status, and for substantial life history forms such as Steelhead and Rainbow Trout. More research is needed to clarify the taxonomy of many groups. In many cases, scientific names are available for different forms, and these are noted.

This guide seeks to facilitate identification of Oregon freshwater fishes. Marine and estuarine fishes are excluded unless they regularly enter freshwater. Knowing the name of an organism is the first step in natural history communication and in understanding an organism's place in nature. The process of identification, even for an experienced professional, can be frustrating. Fishes can differ from each other for reasons beyond their species identity, including age, sex, health, and local environment. Many of our native fishes can also produce hybrids with other species in their family, and these can differ depending on which species was the father and which was the mother. This guide cannot be a comprehensive summary of all these differences but does highlight common hybrids and easily noticed differences due to age and sex.

USING THE GUIDE

The guide is organized into sections. The first section is a guide to fish families and includes a key and a typical example of a species in each family. For each family, there is a short summary, accounts of each species, and sometimes a key. Keys for especially difficult groups include photographs of key characters. Terms are defined in the glossary. The guide can be used to identify fishes by either using the keys or by trying to match a fish to the pictures. In many cases, and especially when first encountering fishes in a new area, it is necessary to use laboratory facilities, such as a dissecting microscope, and comparison to museum specimens to increase confidence in an identification. Proper preservation of specimens can greatly facilitate laboratory examination and the long-term value of the specimen. Good photographs can also be a useful method to document the fishes encountered. Many photographs in this guide are of fishes preserved and deposited in the Oregon State University Ichthyological Collection (OS); the photo voucher specimens are referenced with "OS" plus a catalog number.

SPECIMEN PRESERVATION

The appropriate fish preservation technique depends on your purpose; four are listed below. In all cases, humane practices, such as the AFS-ASIH professional guidelines, should be followed. In addition, good field data should always be collected and a waterproof label included with the specimen to link it unequivocally to the data.

1. *"Life-like" specimens for later examination.* The best procedure is to place anesthetized fish in 10% formalin (5% for larvae) for 3–10 days, soak in water for several days (the formalin smell should be gone), and then store in 50% isopropanol or 70% ethanol. Larger fish should have an incision in the right side, after anesthetization, to accelerate fixation. The bottle with formalin should be large enough that the ratio of the volume of fish to the volume of formalin is at least 1:4. When a specimen is placed in formalin, make sure it lies straight, and try to get fins extended in a normal position. Placing the bottle on its side for 5–10 minutes can help ensure that the specimen fixes straight. Another option, which is more difficult in the field, is to lay the specimen on a wax tray in formalin and use insect pins to

spread out the fins until they are fixed, then put it in a bottle on its side. Wide-mouthed 32-ounce jars with polypropylene lids work well for long-term storage, and similar-sized Nalgene jars work well in the field. There is often confusion about formalin. Purchased 100% formalin is a saturated solution of formaldehyde gas, so it is about 37% formaldehyde. Dilute formalin 1:9 with water to get 10% formalin, which is 3.7% formaldehyde gas. Formalin is a carcinogen, so use gloves and caution.

2. *Genetic samples.* The specimen, or a piece of tissue or fin, can be placed in 95% ethanol or other protective buffer solution, or frozen for storage at -80°C. With ethanol, use a fish-to-volume ratio of 1:5 and use pure ethanol, not denatured ethanol. Also, consider decanting the ethanol and replacing it with fresh pure 95% ethanol after 2 days.

3. *"Life-like" specimens and genetic samples.* Collect a tissue sample before using whole-specimen preservation in formalin.

4. *Otolith aging samples.* The specimen can be placed in 95% ethanol or frozen. If the specimen is large, otoliths can be removed in the field from properly anesthetized or euthanized fish and stored dry in vials or envelopes.

SPECIMEN PHOTOGRAPHY

Good images of fish are difficult to obtain. Drawings, such as the superb illustrations by Joe Tomelleri included herein, are one approach. For those without such talents, photographs offer another challenge. For standardization, photographs and illustrations are of the fish's left side, and cuts or removed fins for genetic samples should be from the right side. There are many problems to overcome to get good color photographs of fish, whether underwater (good examples of northwest fishes can be found at Richard Grost's website) or in a photo tank or aquarium.

By far the most common approach for taxonomic or identification purposes is a narrow vertical photo tank or aquarium where the fish can be "sandwiched" between a movable glass plate and the front of the tank. Key elements are controlling lighting, avoiding glare from the tank, and using a tripod.

An alternative, used for many photographs in this guide, is a horizontal tank and a camera with an antireflection shield. A digital lighthouse—a collapsible white nylon structure—makes an ideal photography environment.

OREGON DRAINAGES

"Fish species are older than the geographic and geologic circumstances in which we find them" (G. R. Smith 1981).

The presence or absence of fish in a body of water is due to geological processes such as mountain building and stream capture, and biological processes such as speciation and extinction. And, of course, humans have confounded patterns with intentional and unintentional transfers. The evolution of Oregon river drainages is complex and not completely understood.

At one scale, the freshwater fishes of Oregon can be conveniently grouped into three faunas. The largest is in the area occupied by the Columbia River and is sometimes called the Cascadia fauna. Within the Columbia fauna, there are at least three recognizable subdivisions: the coastal Tyee, the main stem Columbia, and the Snake.

The second major system is the Klamath River fauna. It has two recognizable subdivisions: the Upper Klamath fauna, and the Rogue River and Lower Klamath fauna. The Rogue River fauna is sparse and seems to have been largely isolated from the coastal Tyee fauna to the north and the Lower Klamath fauna to the south.

In the southeast, a small part of the Lahontan fauna and an area known as the Oregon Lakes seem to have drawn on components of the Columbia, Klamath, and Lahontan faunas. For example, fish in Harney Basin tend to align with Columbia Basin fish while Oregon Lakes basins to the southwest of Harney Basin tend to align with Upper Klamath Basin fish.

In part, the complexity of the Oregon fish fauna is due to the Snake River, which has been connected to both the Columbia and the Klamath Rivers during its long geological history. Because it transported sand from Idaho to the coast of southern Oregon and northern California (Aalto, Sharp, and Renne 1998), we know that a paleo-Snake River crossed the Klamath Mountain province in the mid-Eocene, 66–68 million years ago (Ma). During the Oligocene, about 23–30 Ma, uplift of the Klamath Mountains disrupted this system, and smaller rivers drained the Klamath province (Aalto, Sharp, and Renne 1998). By

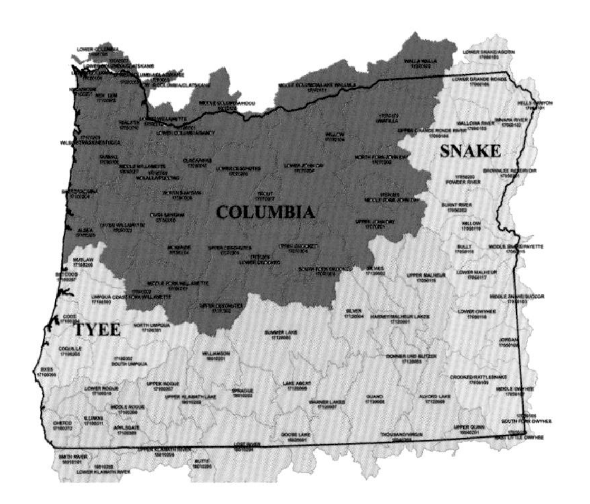
SNAKE
COLUMBIA
TYEE

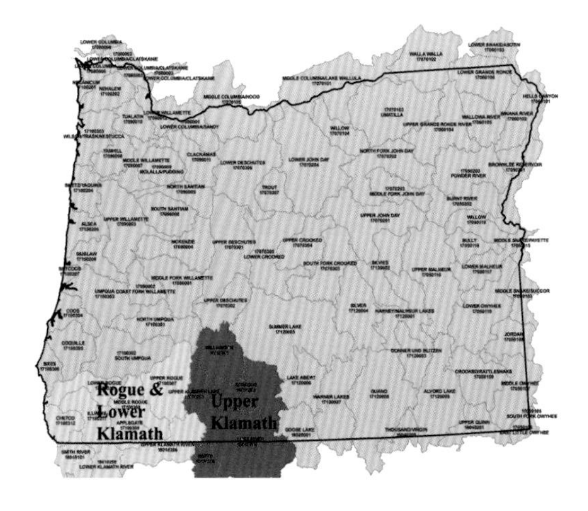
Rogue &
Lower
Klamath
Upper
Klamath

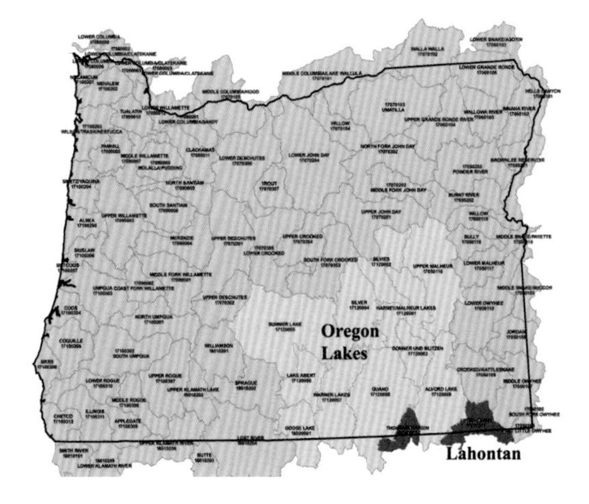
Oregon
Lakes
Lahontan

the late Miocene, 6–9 Ma (Smith 1981), or the early Pliocene, 5 Ma (Aalto, Sharp, and Renne 1998), another paleo-Snake River was re-established across the Klamath province. The Miocene paleo-Snake River drained Lake Idaho during two stages separated by a desiccation event. Similarities in Pliocene fish faunas suggest pre-Pliocene connections between the paleo-Snake, Klamath, and Pit Rivers (Taylor 1985; Smith et al. 2002). About 3 Ma, the paleo-Snake River was captured by the upper Columbia drainage (Smith, Morgan, and Gustafson 2000) in the configuration we see today.

A more recent event of importance to Oregon fishes occurred at the end of the Pleistocene, about 14,000 years ago, when an ice-dammed glacial Lake Missoula in Montana repeatedly discharged huge volumes of water down the Columbia River. The maximum discharge at the point of release, 17 million m^3/second, is the largest known terrestrial freshwater flow. The floods eroded basalt, carving the Channeled Scablands of eastern Washington, and transported sediment as much as 250 km offshore and 800 km south of the Columbia River mouth. The enormity of the events and interesting debates associated with J. Harlan Bretz's discovery of these floods have generated much geophysical research, but little zoogeographic analysis of fish. The Missoula and other Pleistocene floods in the Snake River, such as the Bonneville Floods, had the potential to transport as well as destroy fish faunas, and some of the patterns in fish distribution are likely due to these events. Again, more research is needed to understand the impact of these events on fish distribution.

KEY TO OREGON FISH FAMILIES

1a. No jaws, mouth oval to round and surrounded by papillae; no paired fins; dorsal, caudal, and anal fins continuous; a single median nostril; larvae (ammocoetes) with a small eye spot that is covered over with skin during development; eye forming during abrupt metamorphosis of adult .... Petromyzontidae

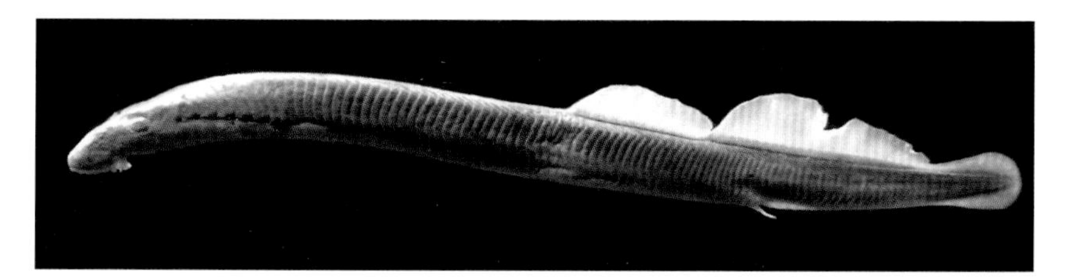

1b. Jaws and paired fins present; nostrils paired; eye developing in egg stage; metamorphosis gradual...................... 2

2a. Caudal fin asymmetrical, with long upper lobe; caudal skeleton heterocercal; sides of body covered with large bony plates... Acipenseridae

2b. Caudal fin symmetrical, with approximately equal upper and lower lobes; caudal skeleton homocercal or slightly heterocercal; sides of body naked or with scales, these occasionally platelike 3

3a. Jaws narrow and elongate, not duck-billed; scales rhomboid; dorsal and anal fins far back on body ... Lepisosteidae

3b. Jaws usually not elongate, but sometimes large and duck-billed; scales present or absent, never rhomboid; dorsal and anal fins variable 4

4a. Pelvic fins abdominal, without spines 5

4b. Pelvic fins subthoracic to thoracic (closer to pectoral fins), with or without spines 17

5a. Dorsal fin with spines (or spinous soft rays); adipose fin present 6

5b. Dorsal fin without spines; adipose fin present or absent 7

6a. Body naked; pectoral fin with a strong, sharp spine; chin barbels present .... Ictaluridae

6b. Body scaled; pectoral fin without strong spines; barbels absent .... Percopsidae

7a. Head without scales 8

7b. Head with scales 14

8a. Midline of belly with modified scales or scutes creating a sawlike profile; no lateral line; adipose eyelid present .... Clupeidae

8b. Midline of belly with a smooth, rounded profile; lateral line usually present; adipose eyelid absent 9

9a. Adipose fin present 10

9b. Adipose fin absent 12

10a. Axillary scales present at base of pelvic fins; young usually with parr marks .... Salmonidae

10b. Axillary scales absent; young without parr marks 11

11a. Body elongate, rounded; jaws with small teeth...........Osmeridae

11b. Body compressed; jaws with large teeth .... Characidae

12a. Barbels present on lower jaw and tip of upper jaw; strong erectile spine below eye .... Cobitidae

12b. Barbels absent or not on upper jaw if present; no erectile spine below eye 13

13a. Ventral profile at anal fin continuous with belly profile; trunk shorter, distance from anal fin origin to caudal fin base more than 50 percent of the distance from anal fin origin to posterior margin of eye; pharyngeal teeth in 1–3 rows with no more than 5 teeth in one row .... Cyprinidae

13b. Ventral profile at anal fin forming obtuse angle with flat belly profile; trunk longer, distance from anal fin origin to caudal fin base less than 50 percent of the distance from anal fin origin to posterior margin of eye; pharyngeal teeth in 1 row of more than 20 teeth .... Catostomidae

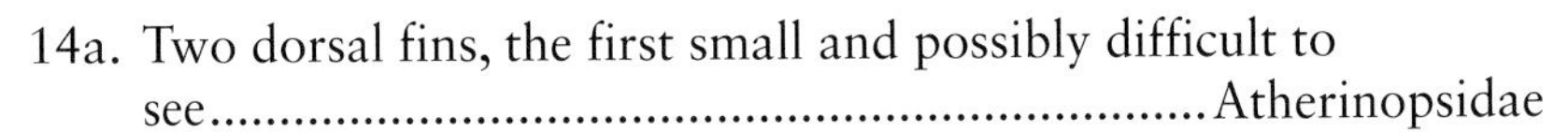

14a. Two dorsal fins, the first small and possibly difficult to see .... Atherinopsidae

14b. One dorsal fin 15

15a. Jaws elongate; duckbill-like snout; mouth terminal; caudal fin forked .... Esocidae

15b. Jaws not elongate; mouth superior; caudal fin rounded 16

16a. Third anal ray unbranched; males with anal fin modified as a gonopodium .... Poeciliidae

16b. Third anal ray branched; male anal fin not modified .... Fundulidae

17a. Dorsal fin with 3 separate spines in front of soft-rayed dorsal fin; body naked or with large lateral scutes .... Gasterosteidae

17b. One or 2 dorsal fins, never 3 separate spines; body scaled 18

18a. Both eyes in adults on 1 side of head; body strongly compressed .... Pleuronectidae

18b. One eye on each side of head; body moderately compressed to rounded 19

19a. Anal fin without spines; barbels on snout and single barbel on chin .... Lotidae

19b. Anal fin with spines; no barbels 20

20a. Anal spines 3 or more 21

20b. Anal spines 1 or 2 22

21a. Opercle with well-developed, posteriorly directed spine .... Moronidae

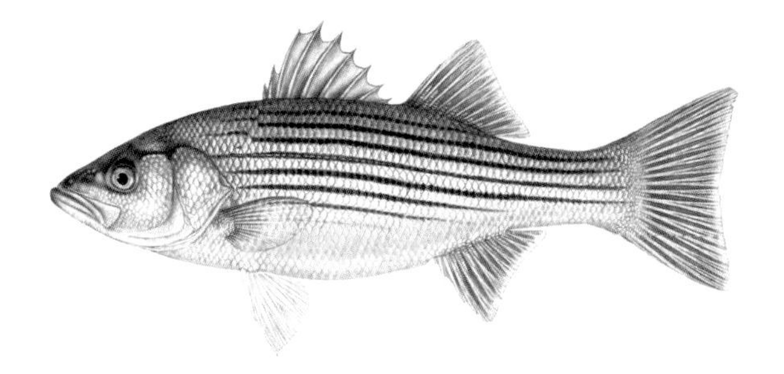

21b. Opercle without well-developed spine but sometimes with posteriorly directed flap .... Centrarchidae

22a. Body naked, sometimes with sandpaper-like prickles; pectoral rays mostly unbranched; head broad, depressed .... Cottidae

22b. Body scaled or naked; pectoral rays mostly branched; head compressed or depressed but not very broad 23

23a. Dorsal fin base with distinctive scaled ridge........... Embiotocidae

23b. Dorsal fin base without scaled ridge.. 24

24a. Dorsal spines stiff and sharp; pelvic fins separated at base .... Percidae

24b. Dorsal spines flexible; pelvic fins united into suction-like disk .... Gobiidae

LAMPREYS
family Petromyzontidae

Lampreys are an ancient lineage of vertebrates lacking many structures, like jaws, gill covers, and paired fins, that evolved in more advanced vertebrates. They have discrete life stages: a larval or ammocoete stage with a rudimentary eye covered by skin, a transformer stage when the eye develops (also called the macrophthalmia stage in anadromous forms), and a reproductive adult stage. The ammocoete, in which the eye spot may be visible in very small specimens, lives many years in soft sediments and feeds on microorganisms, while the adult lives 1 to several years, can be nonfeeding, parasitic, or predatory, and dies after spawning. The dramatic changes in appearance that occur during development and maturation make lamprey identification difficult. Oregon's lamprey diversity is the highest of any state, and the Klamath Basin alone has the most diverse lamprey fauna of any river basin in the world. Morphological and genetic work by Stewart Reid and others suggests there may be even more species.

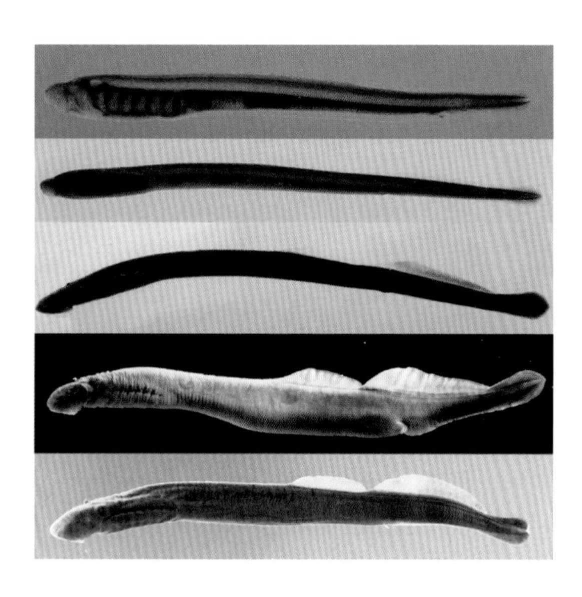

Developmental series of Miller Lake Lamprey. *Top to bottom*: eye-spot ammocoete, 26.2 mm; ammocoete, 39 mm; immature transformed adult, 152 mm; ripe adult female, 96 mm; ripe adult male, 111 mm.

KEY TO GENERA OF AMMOCOETES

1a. Caudal fin dark, uniformly pigmented except for margins, which lack pigmentation; pigmentation on caudal ridge (posterior myomeres) fading from dark anteriad to light posteriad .... *Entosphenus*

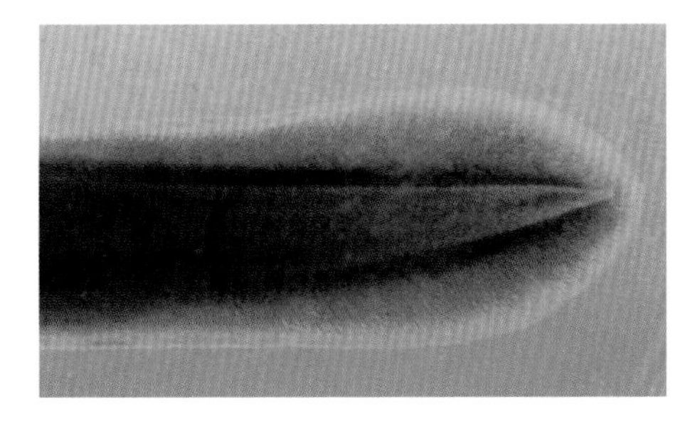

1b. Caudal fin peppered with single melanophores or melanophore aggregations, or melanophores concentrated at caudal fin base and diffuse toward caudal fin margin, which lacks pigmentation, or melanophores completely absent; pigmentation on caudal ridge uniformly dark .... *Lampetra*

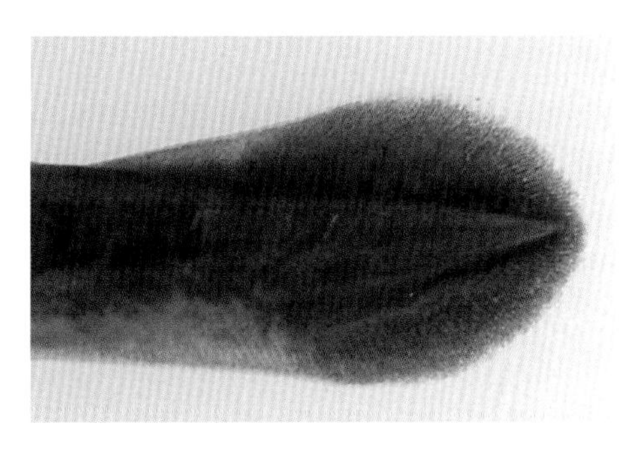

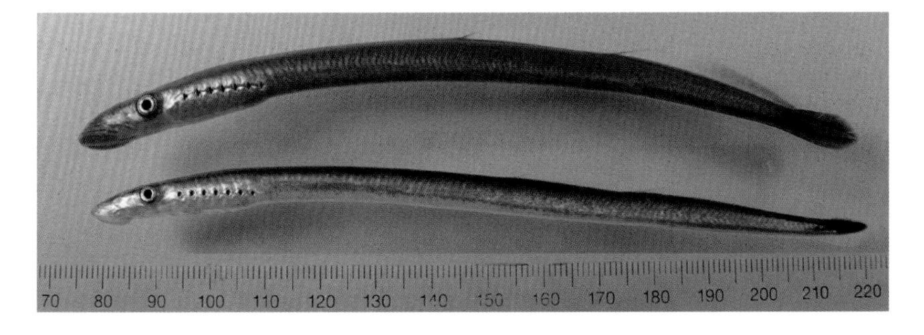

Caudal pigmentation in transformers is similar to that of ammocoetes above. Above, Entosphenus macrophthalmia stage; below, Lampetra transformer. Photo credit, Greg Kovalchuk.

KEY TO SPECIES OF ADULTS (EYE PRESENT)

1a. Total length greater than 330 mm .... *E. tridentatus*

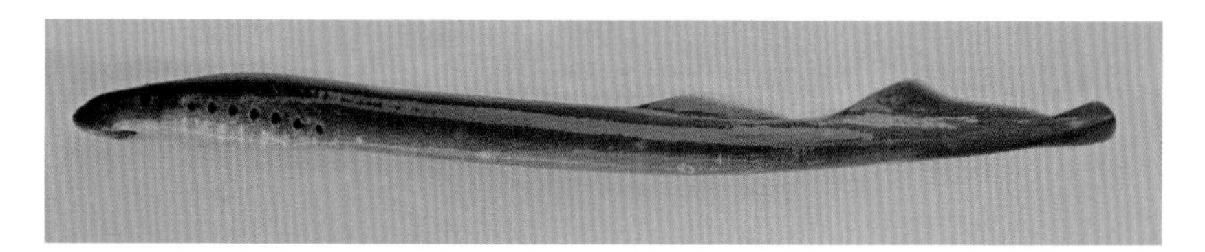

Pacific Lamprey, *Entosphenus tridentatus* (Richardson, ex Gairdner, 1837). Adult, 580 mm TL, Umpqua River, 12 July 2013 (released). Anadromous and parasitic, widespread. Two interior resident forms are thought to be separate species, the Klamath Lake Lamprey, *Entosphenus* sp. 1, and the Goose Lake Lamprey, *Entosphenus* sp. 2.

1b. Total length less than 330 mm 2

2a. Anadromous; eye moderate to large, diameter about equal to or slightly less than distance from posterior edge of eye to anterior edge of first branchial pore................3

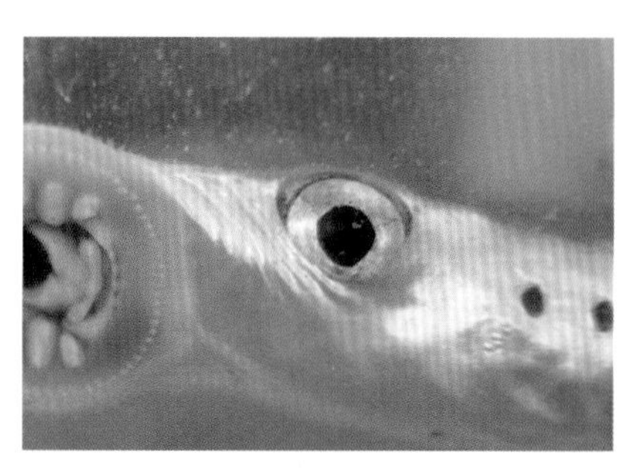

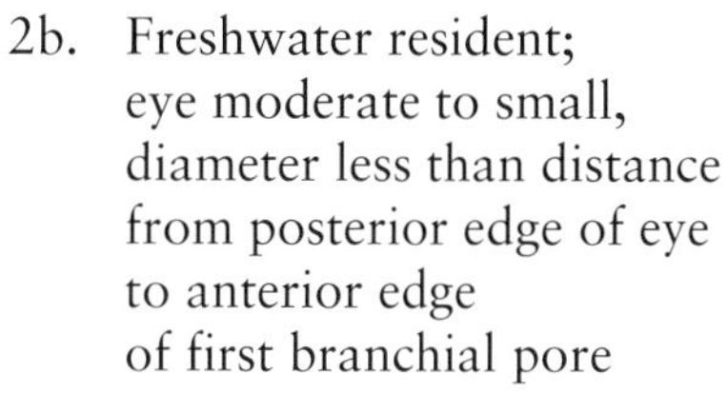

2b. Freshwater resident; eye moderate to small, diameter less than distance from posterior edge of eye to anterior edge of first branchial pore

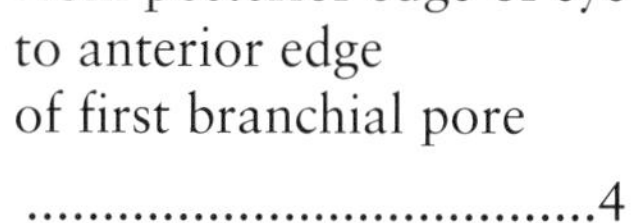

....4

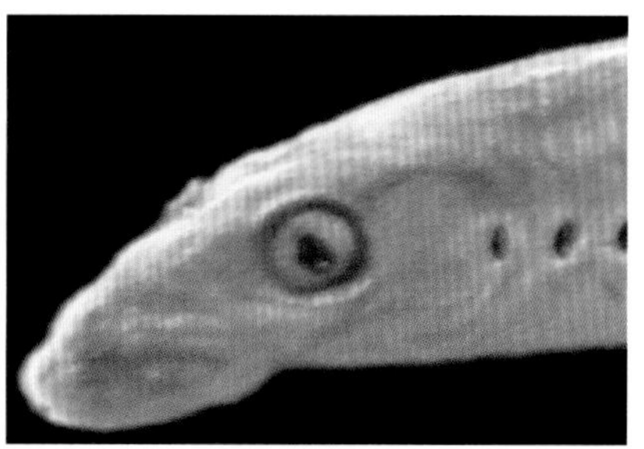

3a. Supraoral with 3 cusps, infraoral with 5–6 cusps, 4 inner laterals; caudal fin darkly pigmented throughout; generally dark blue gray dorsally, with silvery sides and belly; iris light blue to silvery in live specimens; metamorphosed juveniles in freshwater about 100–160 mm TL .... *E. tridentatus* (see above).

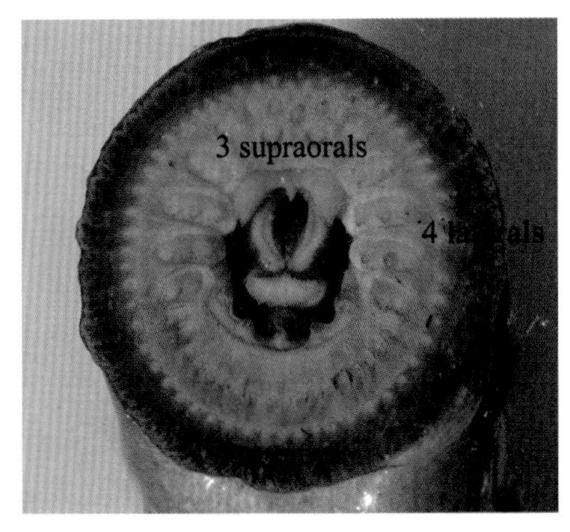

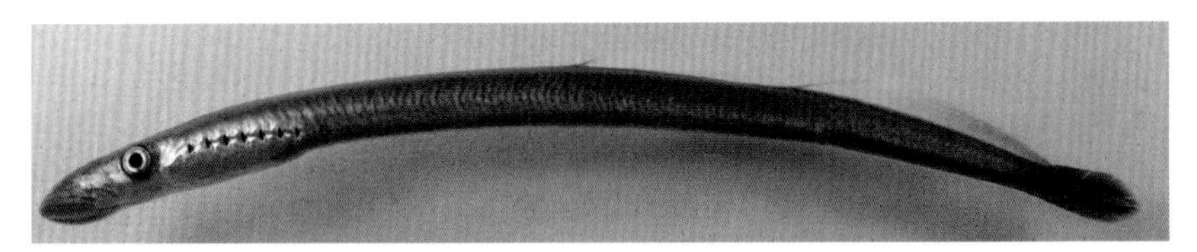

Metamorphosed juvenile (macrophthalmia) Pacific Lamprey, about 145 mm, John Day Dam. Photo credit, Greg Kovalchuk.

3b. Supraoral with 2 cusps, infraoral with 7+ cusps, 3 inner laterals; caudal fin melanophores concentrated at fin base and diffuse toward fin margin, which lacks pigmentation, dorsal fin unpigmented except perhaps in spawners; generally dark dorsally, with light belly; metamorphosed juveniles and adults 115–315 mm TL ....*L. ayresii*

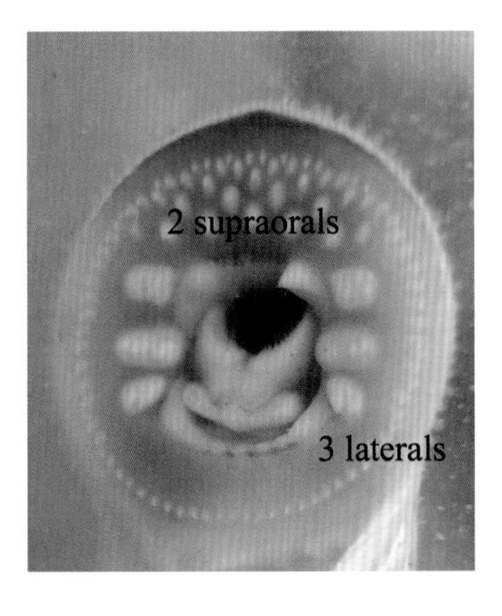

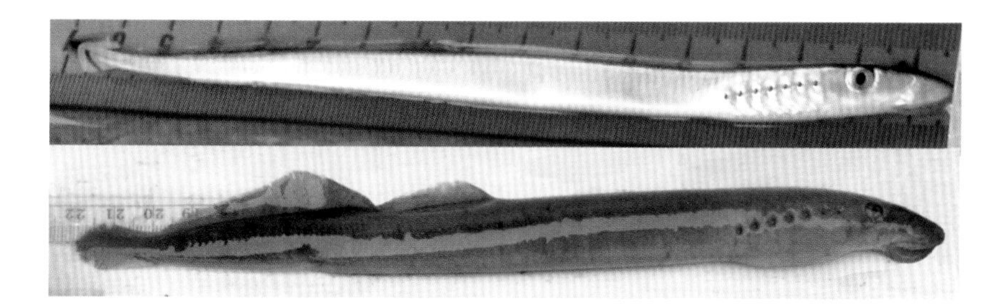

Western River Lamprey, *Lampetra ayresii* (Gunther, 1870). Parasitic, anadromous, Lower Columbia and central coastal rivers of Oregon. Part of the Brook Lamprey complex. *Above*, immature, about 170 mm, Columbia River. Photo credit, Laurie Weitkamp. *Below*, mature adult male, about 220 mm, Puget Sound, Washington. Photos of mouth and adult by Mike Hayes.

4a. Oral disk large, diameter greater than half the branchial length 5

4b. Oral disk small, diameter less than half the branchial length..................................8

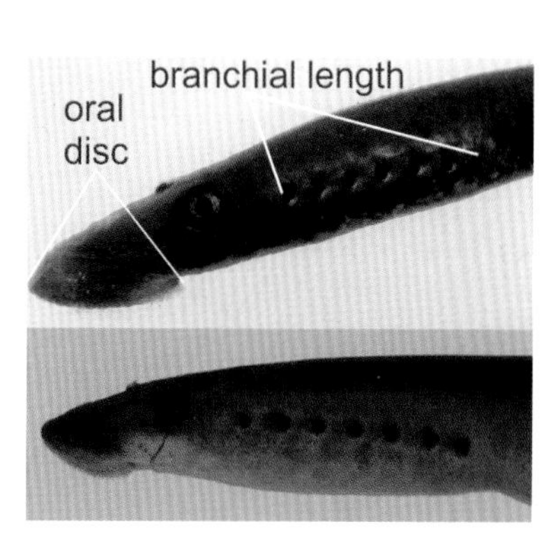

5a. Sexually mature, both sexes with expanded second dorsal, females with pseudoanal and cloacal swelling, males with protruding papilla 6

5b. Immature, no secondary sexual characters
.... juveniles of E. similis, E. minimus, or E. sp. ("Klamath Lake" Łamprey)

6a. Mature adults > 150 mm 7

6b. Mature adults 75–130 mm..*E. minimus*

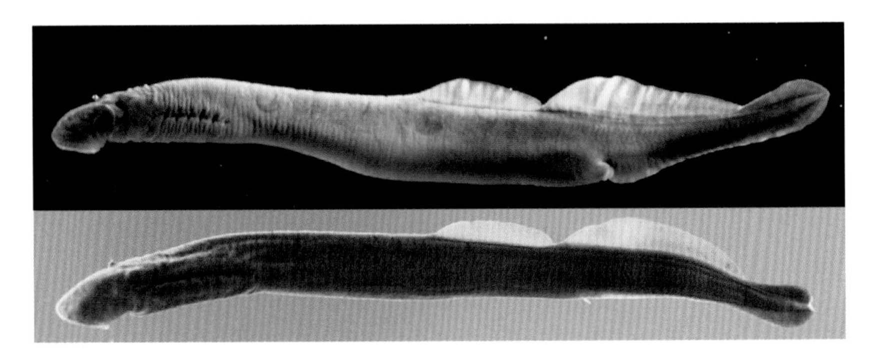

Miller Lake Lamprey, *Entosphenus minimus* (Bond & Kan, 1973). *Above*, ripe female with spawning scars, 96 mm, OS 15866; *below*, ripe male, 111 mm, OS 16844. Small (mature to 126 mm, immature transformers to 163 mm), predatory, Klamath Basin, primarily in upper basin but downstream extent uncertain.

7a. Disk small, 5.6%–8.4% of TL, prebranchial length 9.2%–10.2% of TL; up to 315 mm TL; infraoral lamina with more than 5 teeth; 3 laterals on each side of the oral disk .... *E. tridentatus* complex
("Klamath Lake" Lamprey or "kawaiga")

7b. Disk large, 7.8%–10.4% of TL, prebranchial length 13.0%–16.4% of TL; up to ca. 270 mm TL, infraoral lamina usually with 5 teeth; typically 4 laterals on each side of the oral disk .... *E. similis*

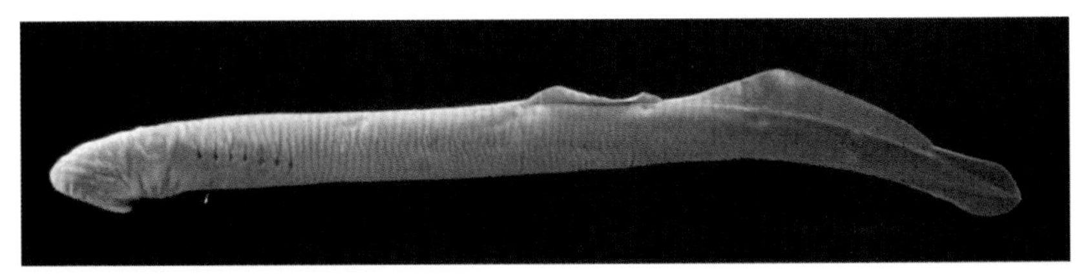

Klamath Lamprey, *Entosphenus similis* Vladykov & Kott, 1979. Adult, 155 mm, OS 13717, Spencer Creek, Klamath River. Parasitic, resident and endemic in upper and lower Klamath basins.

8a. Caudal fin darkly pigmented; supraoral with 3 cusps (often weak), infraoral with 5–6 cusps, 4 inner laterals; dorsal body coloration uniform .... *E. lethophagus*

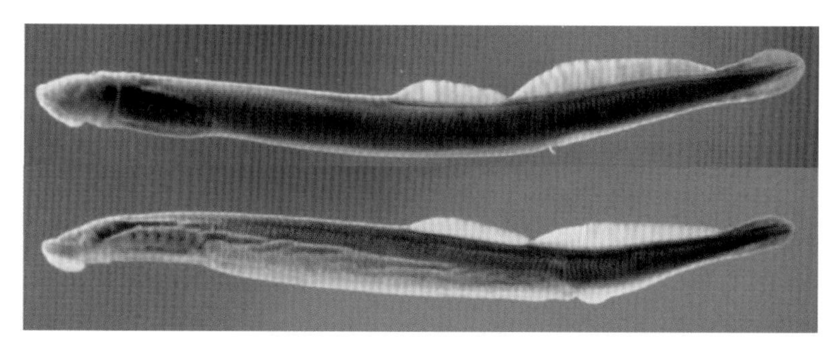

Pit-Klamath Brook Lamprey, *Entosphenus lethophagus* Hubbs, 1971. *Above*, adult male, 133 mm, OS 16790; *below*, adult female, 184 mm, OS 16791, both from Big Springs Creek. Nonparasitic, Upper Klamath Basin. A related species, the Modoc Brook Lamprey, *Entosphenus folletti* Vladykov & Kott, 1976, may also occur in Oregon.

8b. Caudal fin unpigmented; supraoral tooth plate with 2 separate cusps, infraoral with 7–8 cusps, 3 inner laterals; body often mottled dorsally 9

9a. Trunk myomeres 62–69; papillae in the third gill pore 15–32; caudal fin angular or slightly pointed ....*L. richardsoni*

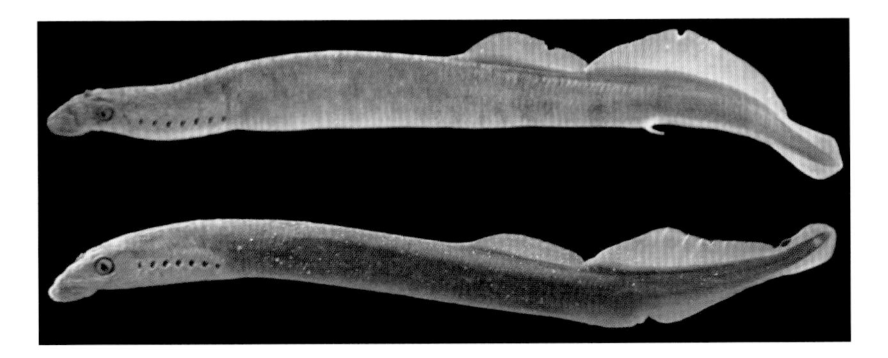

Western Brook Lamprey, *Lampetra richardsoni* Vladykov & Follet, 1965. *Above*, male, 125 mm; *below*, female, 119 mm, both OS 17967, Coquille River. Nonparasitic, more coastal-oriented form than *L. pacifica*, but distribution uncertain. Part of a complex that includes River Lamprey, Pacific Brook Lamprey, and possibly several undescribed cryptic species.

9b. Trunk myomeres 55–63; papillae in the third gill pore 6–21; caudal fin rounded ....*L. pacifica*

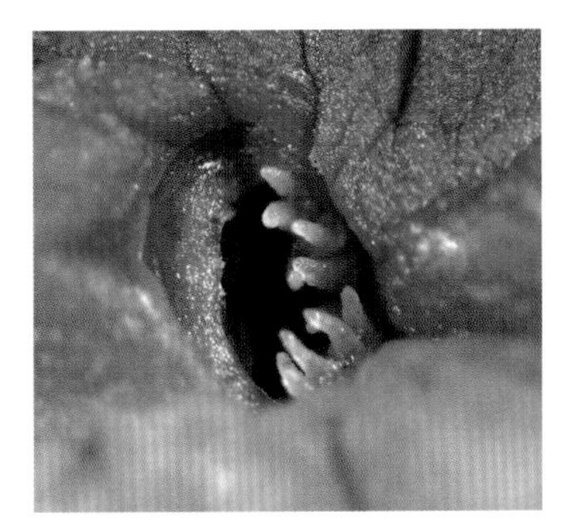

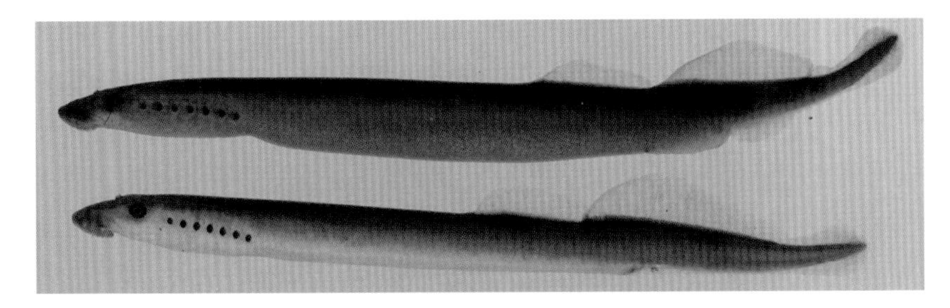

Pacific Brook Lamprey, *Lampetra pacifica* Vladykov, 1973. *Above*, female, 146 mm; *below*, male, 135 mm, both OS 18405, Clatskanie River. Nonparasitic, western Oregon, distribution uncertain.

STURGEONS
family Acipenseridae

Sturgeons include some of the largest freshwater fishes in the world. As the original source of caviar and a popular sport fish, they are overexploited in many places. In Oregon, sturgeon decline is often attributed to environmental changes.

The White Sturgeon, *Acipenser transmontanus*, is our largest freshwater fish, reaching sizes of up to 19 ft. and 1,800 lb. (5.8 m and 816 kg). It is also our longest lived, with estimated ages of up to 100 years. The smaller Green Sturgeon, *A. medirostris*, reaches lengths of about 7 ft. (2.1 m), weighs up to 350 lb. (159 kg), and lives up to 70 years. Adult Green Sturgeon are marine but spawn in freshwater, while White Sturgeon can complete their whole life in freshwater.

The White Sturgeon has a short, rounded rostrum, with barbels closer to the rostrum than to the mouth, and numerous (38–48) lateral plates. The Green Sturgeon has a longer, narrow rostrum, with barbels closer to the mouth, and fewer (23–30) lateral plates.

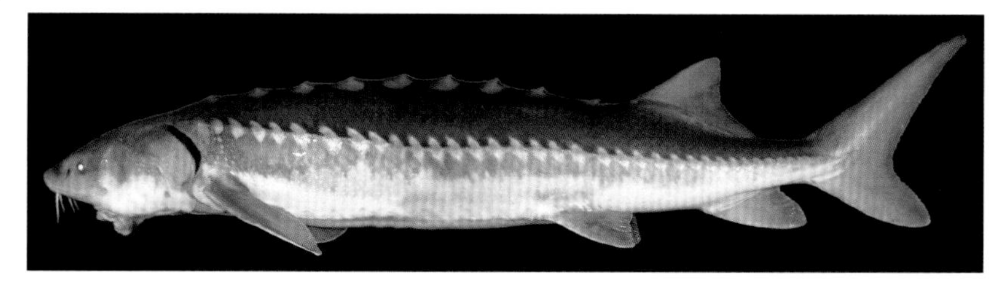

White Sturgeon, *Acipenser transmontanus* Richardson, 1837. 100 cm TL. Photo credit, René Reyes. Anadromous, Columbia and Snake Rivers.

Green Sturgeon, *Acipenser medirostris* Ayres, 1854. Illustration by Joe Tomelleri. Anadromous, widespread along coast.

GARS
family Lepisosteidae

The gars are an eastern North American family of fishes. The carcass of a single Florida Gar, *Lepisosteus platyrhincus*, was collected in 1999 by Dan Gurr on the South Yamhill River (OS 17180). No viable populations are known in Oregon.

Florida Gar, *Lepisosteus platyrhincus* DeKay, 1842. Illustration by Joe Tomelleri.

HERRINGS
family Clupeidae

American Shad, *Alosa sapidissima*, are native to the eastern United States and were introduced to the Sacramento River in 1871 and the Willamette River in 1885. Several million are frequently counted at Bonneville Dam. They are found in larger coastal streams and the Columbia River.

American Shad are laterally compressed, with deciduous fine scales and a series of dark spots on the flank behind the head. The adults have modified scales on the belly, called scutes, that are sharp ("saw belly") and give rigidity to the belly.

American Shad, *Alosa sapidissima* (Wilson, 1811). Photo credit, René Reyes. Anadromous, Columbia and coastal rivers.

MINNOWS
family Cyprinidae

The minnows are the largest family of vertebrates, with well over 2,000 species, and the largest family of fishes in Oregon, with both native and nonnative species. Their jaws are toothless but they have well-developed teeth on pharyngeal jaws in the throat.

All of our native minnows are sexually dimorphic, with males having longer pectoral fins so there is a smaller gap between the tip of the pectoral fin and the origin of the pelvic fin. When sexually active, all minnows have breeding tubercles (small bumps) on fins and scales; these are usually more prominent on males but can be found on both sexes. Spawning usually occurs in late spring and summer. Larval and young minnows can be seen in large numbers in shallow water along shorelines of lakes and streams in early summer.

Confident identification can require microscopic examination of scales, pharyngeal teeth, and small structures.

Willamette Black-Lined Dace. *Top*, juvenile, 21.6 mm (OS 18426); *middle*, spawning male, 48.4 mm (OS 18452); *bottom*, spawning female, 65.3 mm (OS 18452).

Hybridization (including crosses between genera), as in this Speckled Dace × Redside Shiner (OS 18701), is common in minnows.

KEY TO CYPRINIDAE

1a. Dorsal fin long, base longer than head, more than 12 rays; anterior spine (spinous soft rays) in dorsal and anal fins with serrations on posterior edge 2

1b. Dorsal fin shorter, base shorter than head, fewer than 12 rays; all fins with soft rays (leading rays may be hard, but no serrated spines on fins) 3

2a. Upper jaws without barbels; anterior tip of upper jaw level with middle of eye .... *Carassius auratus*

Goldfish, *Carassius auratus* (Linnaeus, 1758). *Above*, 10 mm, KR03198; *below*, 64 mm, OS 18412, Klamath River. Introduced and widespread.

2b. Upper jaws with 2 pairs of barbels; anterior tip of upper jaw below eye .... *Cyprinus carpio*

Common Carp, *Cyprinus carpio* Linnaeus, 1758. 440 mm, Harney Basin. Introduced and widespread.

3a. Numerous fine scales (> 100 in lateral line); large maxillary barbel; rounded fins with slightly emarginate caudal; iris red .... *Tinca tinca*

Tench, *Tinca tinca* (Linnaeus, 1758). 125 mm, OS 9119. Introduced, mostly upper Columbia River.

3b. Scales large or small but always fewer than 100 in lateral line; maxillary barbel present or absent; fin shapes variable, caudal rounded to deeply forked; iris not red 4

4a. Anal fin origin distinctly behind posterior tips of depressed dorsal fin rays 5

4b. Anal fin origin slightly behind to well forward of depressed posterior tips of dorsal fin rays 6

5a. Small barbels at corners of mouth visible with magnification; scales small, 66–86 in lateral line; eyes laterally placed ....*Mylocheilus caurinus*

Peamouth, *Mylocheilus caurinus* (Richardson, 1836). *Above*, 78 mm, OS 18444, Oswego Creek; *below*, 195 mm, OS 18407, Milton Creek. Columbia Basin endemic.

5b. No barbels; scales large, about 34–37 in lateral line; downward-directed eyes ....*Ctenopharyngodon idella*

Grass Carp, *Ctenopharyngodon idella* (Valenciennes, 1844). Illustration by Joe Tomelleri. Introduced legally in Devils Lake, but now in Willamette and lower Columbia Rivers.

6a. Lower jaw with straight-edged hard cartilaginous plate; caudal peduncle slightly "pinched," with lower procurrent caudal rays creating an abrupt transition to the caudal fin .... *Acrocheilus alutaceus*

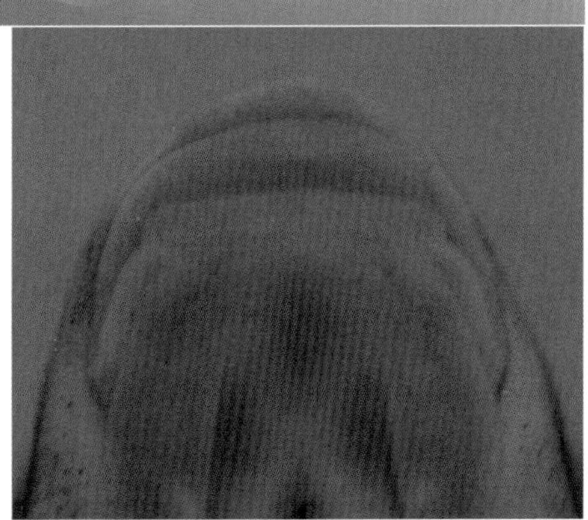

Chiselmouth, *Acrocheilus alutaceus* Agassiz & Pickering, 1855. *Top*, juvenile, 21.5 mm; *middle*, 125 mm, OS 17002; *bottom*, "chisel" jaw. Columbia Basin endemic. For hybrids, see Northern Pikeminnow below.

6b. Lower jaw normal, without straight chisel-like plate; lower procurrent caudal rays rising gradually from caudal peduncle 7

7a. Scales large, with a salt-and-pepper pattern, fewer than 40 in lateral line; caudal peduncle depth shallow 8

7b. Scales small, with or without salt-and-pepper pattern, usually more than 50 in lateral line 9

8a. Anal fin origin about directly under dorsal fin insertion; breast usually unscaled, only 7% of specimens with over half of breast scaled .... *Oregonichthys crameri*

Oregon Chub, *Oregonichthys crameri* (Snyder, 1908). 38 mm, OS 12077. Willamette River endemic.

8b. Anal fin origin distinctly in front of dorsal fin insertion; breast usually scaled, about 95% with breast completely scaled .... *Oregonichthys kalawatseti*

Umpqua Chub, *Oregonichthys kalawatseti* Markle, Pearsons & Bills, 1991. 37 mm, OS 17893. Umpqua River endemic.

9a. Mouth large, upper jaw reaching or extending past a vertical through anterior margin of eye; usually 9 dorsal rays 10

9b. Mouth smaller, upper jaw not reaching a vertical through anterior margin of eye; usually 6–9 dorsal rays......... 11

10a. Number of transverse scales from dorsal fin origin counting posteriad to lateral line 15–18; number of circumpeduncular scales around narrowest part of caudal peduncle 27–30; caudal spot prominent, especially in young .... *Ptychocheilus oregonensis*

Northern Pikeminnow, *Ptychocheilus oregonensis* (Richardson, 1837). *Above*, juvenile, 21 mm; *below*, adult, no data. Columbia Basin endemic. Hybridizes frequently.

Above, hybrid Chiselmouth × Pikeminnow, 140 mm, OS 12471; *below, left to right*, jaws of Chiselmouth, hybrid, Northern Pikeminnow.

10b. Number of transverse scales from dorsal fin origin counting down and back to lateral line 19–23; number of circumpeduncular scales around narrowest part of caudal peduncle 32–36 .... *Ptychocheilus umpquae*

Umpqua Pikeminnow, *Ptychocheilus umpquae* Snyder, 1908. 295 mm, OS 15462. Tyee Basin endemic, introduced in Rogue River. There is some evidence that the Siuslaw River form might be taxonomically different.

11a. Abdomen with fleshy, scaleless keel between pelvic fins and anal fin; pelvic fin with axillary scale at base; body deep, with strongly decurved lateral line .... *Notemigonus crysoleucas*

Golden Shiner, *Notemigonus crysoleucas* (Mitchill, 1814). 85 mm, OS 19347, Diamond Lake. Introduced widely, in Oregon west of Cascades.

11b. Abdomen without fleshy, scaleless keel between pelvic fins and anal fin; axillary scale at base of pelvic fin present or absent; body shape variable, if deep, lateral line moderately decurved 12

12a. Short, procurrent dorsal ray well separated from first primary, unbranched dorsal fin ray and connected by a membrane, creating a discontinuous outline (prominent in males); mature females with greatly reduced anal fin; mature males with prominent tubercles on snout and spongy mass between head and dorsal fin origin .... *Pimephales promelas*

Fathead Minnow, *Pimephales promelas* Rafinesque, 1820. *Top*, juvenile, 34 mm, OS 18411; *middle*, female, 43 mm, OS 14116; *bottom*, male, 47 mm, OS 14116. Introduced, scattered localities, abundant in Klamath Basin.

12b. Short, procurrent dorsal ray tightly applied to front of first primary, unbranched dorsal fin ray, creating a continuous outline; mature females without reduced anal fin; mature males without spongy mass between head and dorsal fin origin.. 13

13a. Body elongate, dorsal and ventral profiles of caudal peduncle subparallel; maxillary barbels usually present; mouth terminal to subterminal 14

13b. Body more robust, caudal peduncle tapering to caudal fin; maxillary barbels absent; mouth terminal 21

14a. Upper jaw not protrusible, with frenum attaching it to skull; mouth subterminal 15

14b. Upper jaw protrusible, with groove around snout; mouth terminal to subterminal... 17

15a. Dorsal rays 9–10; anal fin origin in front of vertical from base of last dorsal fin ray; dorsal fin origin over or in front of vertical from base of last pelvic fin ray .... *Rhinichthys evermanni*

Umpqua Dace, *Rhinichthys evermanni* Snyder, 1908. Male, 107 mm, West Fork Smith River. Umpqua River endemic. Photo credit, Bruce Miller.

15b. Dorsal rays 8; anal fin origin behind vertical from base of last dorsal fin ray; dorsal fin origin over or behind vertical from base of last pelvic fin ray 16

16a. Caudal peduncle slender; usually 30 (29–31) scales around caudal peduncle; caudal fin with shallow fork, its apex in the posterior third of the caudal fin ....*Rhinichthys* sp. (Millicoma Dace)

Millicoma Dace, *Rhinichthys* sp. (see Bisson & Reimers, 1977, *Copeia*, 518–22). Male, 77.4 mm, OS 19083. Coos River endemic. More research is needed on morphological differences with Umpqua and Longnose Daces.

16b. Caudal peduncle slightly deeper; usually 28 (25–30) scales around caudal peduncle; caudal fin with deeper fork, its apex near the middle of the caudal fin .... *Rhinichthys cataractae*

Longnose Dace, *Rhinichthys cataractae* (Valenciennes, 1842). *Above*, 30 mm, OS 18432, Willamette River; *below*, male, 92 mm, OS 16967, South Santiam River. Widespread in North America, Columbia Basin in Oregon.

17a. Scales small, about 70–80 in lateral line; snout barely overhanging upper jaw; border above belly below lateral line an irregular, wavy margin trending upward from anal fin.................................... *Rhinichthys osculus klamathensis*

Western Speckled Dace, *Rhinichthys osculus klamathensis* (Cope, 1872). *Top*, male, 56 mm, OS 15863, Barnes Valley Creek; *middle*, female, 51 mm, OS 15852, Foskett Spring; *bottom*, juvenile, 22 mm, OS 18044, Quinn River. Widespread form(s) in southeastern Oregon, but more research is needed.

17b. Scales larger, about 50–70 in lateral line; snout variable; secondary lateral stripe variable, lower margin above belly not wavy 18

18a. Dorsal rays usually 9; part of the shaft of the last pelvic ray attached to the side of the body by membrane, other membranous stays possibly attached to bases of other pelvic rays 19

18b. Dorsal rays usually 7–8; pelvic rays not attached to side of body by membranous stays 20

19a. Anal fin origin forward, in front of vertical from base of last dorsal ray; fork in caudal fin shallow, apex in posterior third of fin; pelvic fin rays without membranous stays; dorsal and anal fins rounded, distal section of rays similar in coloration to basal section; body deep, robust; maxillary barbel short, not protruding beyond maxillary groove; some scales darkened, giving a freckled appearance...... *Rhinichthys umatilla*

Umatilla Dace, *Rhinichthys umatilla* (Gilbert & Evermann, 1894). *Top*, male, 56 mm; *middle*, female, 59 mm; *bottom*, juvenile, 30 mm, all OS 18283, Umatilla River. Middle and upper Columbia, distribution uncertain.

19b. Anal fin origin under vertical from base of last dorsal ray; fork in caudal fin deep, apex near middle of fin; bases of last several pelvic fin rays usually attached to body by fleshy membranes (stays), last pelvic ray always attached up to half its length to body; dorsal and anal fins falcate, distal section of rays lighter than basal section; body elongate, slender; maxillary barbel long, protruding beyond corner of mouth; sides with irregular blotches covering multiple scales.............................. *Rhinichthys falcatus*

Leopard Dace, *Rhinichthys falcatus* (Eigenmann & Eigenmann, 1893). *Above*, 24 mm, OS 18433; *below*, male, 62 mm, OS 17185, both Willamette River. Columbia Basin, scattered (Willamette, Umatilla, Snake). The form in the Snake River has a deeper caudal peduncle.

20a. Mouth terminal; barbels small or absent; 7–8 dorsal rays; lateral line complete; posterior tip of last dorsal ray over or in front of vertical from base of first anal ray .... *Rhinichthys osculus* ssp.

Malheur Spring Dace, *Rhinichthys osculus* ssp. Female, 39 mm, OS 15851, Barnyard Spring. Perhaps a Malheur Basin endemic, but more research is needed on distribution and taxonomic status.

20b. Mouth subterminal; barbels small, usually present; usually 8 dorsal rays; lateral line incomplete; posterior tip of last dorsal ray over or behind vertical from base of last anal ray; snout long, extending past margin of upper jaw; dark lateral band on side continuing through eye .... *Rhinichthys osculus nubilus*

Black-Lined Speckled Dace, *Rhinichthys osculus nubilus* (Girard, 1856). Male, 46 mm, OS 17968, Coquille River. Widespread form in western Oregon, but more research is needed on distribution and taxonomic status.

21a. Pelvic axillary scale present; 8–20 anal fin rays; side with dark lateral band bordered dorsally by a thinner yellow band that extends to posterior margin of orbit; spawners with a red wash on side below lateral line 22

21b. Pelvic axillary scale absent; 6–9 anal fin rays; side with or without dark lateral band and not bordered by a colored band or wash; spawners with uniform bronze to dark colors on side 25

22a. Anal fin origin behind a vertical through base of last dorsal fin ray; anal fin base short, with 8–10 rays .... *Richardsonius egregius*

Lahontan Redside Shiner, *Richardsonius egregius* (Girard, 1858). 62 mm, OS 18050, McDermitt Creek. Endemic to Lahontan Basin, in Quinn River in Oregon.

22b. Anal fin origin under or in front of a vertical through base of last dorsal fin ray; anal fin base variable, with 10–20 rays 23

23a. Anal rays usually14–18 (range 13–22); body deep, in specimens > 75 mm SL, dorsal origin to anal origin distance 109%–128% of head length; spawners with yellow cheeks......................... *Richardsonius balteatus balteatus*

Columbia Redside Shiner, *Richardsonius balteatus balteatus* (Richardson, 1837). *Above*, 15 mm, OS 18064, Succor Creek; *below*, female, 88 mm, OS 15472, Owyhee River. Widespread in Columbia Basin.

23b. Anal rays usually 10–15; body depth relatively shallow, in specimens > 75 mm SL, dorsal origin to anal origin distance 86%–110% of head length; spawners with red or yellow cheeks... 24

24a. Spawners with red to orange blush on cheeks; anal rays usually 12–13 (range 10–15) .... *Richardsonius balteatus siuslawi*

Tyee Redside Shiner, *Richardsonius balteatus siuslawi* (Evermann & Meek, 1898). Male, 69 mm, OS 16849, Siuslaw River. Coastal streams of Tyee drainages, perhaps in parts of Willamette River, but more research is needed to determine whether this is a separate species.

24b. Spawners with yellow cheeks; anal rays usually 11–12 (range 10–13) .... *Richardsonius balteatus hydrophlox*

Bonneville Redside Shiner, *Richardsonius balteatus hydrophlox* (Cope, 1872). Male, 87 mm, OS 16845, Oak Creek, Willamette River. Widespread in Columbia Basin, but more research is needed on taxonomic status and coexistence with Columbia Redside Shiner.

25a. Scales with radii in anterior and posterior fields 26

25b. Scales with radii in posterior field only.................................... 28

26a. Fewer than 10 gill rakers; usually 8 (7–10) dorsal fin rays .... *Hesperoleucus symmetricus*

California Roach, *Hesperoleucus symmetricus* (Baird & Girard, 1854). 60 mm, OS 18070. Goose Lake.

26b. Gill rakers 13–22; usually 7 (6–8) dorsal fin rays; Alvord Basin 27

27a. Lateral line in specimens > 55 mm typically < 30% complete, with fewer than 20 pored scales (2 large specimens with > 60 scales); upper jaw length usually reaching to vertical from front margin of eye and about equal (77%–116%) to caudal peduncle depth; at similar sizes, snout larger; modal number of pectoral rays 13 (range 12–15)... *Siphateles boraxobius*

Borax Lake Chub, *Siphateles boraxobius* Williams & Bond, 1980. 66 mm, OS 18033. Endemic to Borax Lake. Photo credit, Travis Neal.

27b. Lateral line in specimens > 55 mm typically > 40% complete, with more than 40 pored scales; upper jaw length usually not reaching to vertical from front margin of eye and much less (54%–78%) than caudal peduncle depth; at similar sizes, snout smaller; modal number of pectoral rays 14 (range 12–16).................. *Siphateles alvordensis*

Alvord Chub, *Siphateles alvordensis* Hubbs & Miller, 1972. 44 mm, OS 18039. Endemic to Alvord Basin in Oregon and Nevada. Photo credit, Travis Neal.

28a. Gill rakers 16–24 29

28b. Gill rakers 9–16 30

29a. Body and head compressed, head pointed, jaw moderately to very oblique; ventral profile from angle of lower jaw to pelvic fin essentially flat; lateral abdominal scale pockets not outlined by melanophores (Oregon populations only) .... *Siphateles bicolor thalassinus*

Warner Tui Chub, *Siphateles bicolor thalassinus* (Cope, 1883). 68 mm, OS 15820. In Goose Lake and Warner Valley and introduced in Summer Lake Basin. More research is needed on taxonomic status.

29b. Body and head wide, head rounded, jaw slightly oblique; ventral profile from angle of lower jaw to pelvic rounded; lateral abdominal scale pockets outlined by melanophores .... *Siphateles bicolor eurysoma*

Sheldon Tui Chub, *Siphateles bicolor eurysoma* (Williams & Bond, 1981). 115 mm, OS 15854. In Guano and Catlow Valleys of Oregon and Nevada. More research is needed on taxonomic status.

30a. Pelvic fin origin under or slightly in front of dorsal fin origin; caudal peduncle elongate; gill rakers 9; spawning males with blue snout............................*Gila coerulea*

Blue Chub, *Gila coerulea* (Girard, 1856). *Top*, 14 mm, A09332; *middle*, 46 mm, OS 18414; *bottom*, 116 mm, OS 18417, all Upper Klamath Lake. Also known as *Klamathella coerulea*. Endemic to Klamath Basin of Oregon and California.

30b. Pelvic fin origin behind dorsal fin origin; caudal peduncle deeper; gill rakers 9–24; snout never blue 31

31a. Lateral line scales usually < 50 (41–54); caudal peduncle somewhat elongate, its length more than twice its depth........ 32

31b. Lateral line scales usually > 50 (45–65); caudal peduncle short, its length less than twice its depth 33

32a. Scales from dorsal origin to lateral line 7–10; modal number of dorsal rays 8 .... *Siphateles bicolor bicolor*

Tui Chub, *Siphateles bicolor bicolor* (Girard, 1856). *Above*, 16 mm, A09198; *below*, 123 mm, OS 18416, both Upper Klamath Lake. Klamath Basin or Great Basin endemic, but more research is needed on taxonomic status.

32b. Scales from dorsal origin to lateral line 11–13; modal number of dorsal rays 9 .... *Siphateles bicolor columbianus*

Columbia Tui Chub, *Siphateles bicolor columbianus* (Snyder, 1908). 70 mm, OS 15842. Upper Columbia of Washington and Oregon, including Malheur Basin. More research is needed on taxonomic status.

33a. Body elongate; belly dark dusky, lateral scale pockets outlined with melanophores, occasionally creating diamond patterns .... *Siphateles bicolor oregonensis*

Oregon Lakes Tui Chub, *Siphateles bicolor oregonensis* (Snyder, 1908). *Above*, female, 80 mm; *below*, male, both OS 18279, XL Spring. May be aligned with *S. b. obesus*, but more research is needed on taxonomic status. In Lake Abert, Summer Lake, Silver Lake, and Hutton Springs of south-central Oregon.

33b. Body robust, deep; belly pale, lateral scale pockets not outlined with melanophores.. 34

34a. Usually 8 dorsal rays; pharyngeal teeth in 1 row; restricted to Quinn River drainage in Oregon ....*Siphateles bicolor obesus*

Lahontan Tui Chub, *Siphateles bicolor obesus* (Girard, 1856). 132 mm, OS 18029, Nevada-Oregon border. In Oregon only in Quinn River, but it may now be extinct in Oregon. More research is needed on taxonomic status.

34b. Usually 9 dorsal rays; pharyngeal teeth in 2 rows; introduced in Owyhee River .... *Gila atraria*

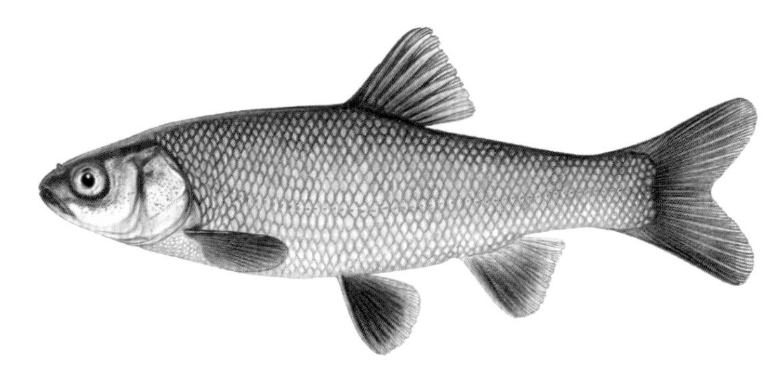

Utah Chub, *Gila atraria* (Girard, 1856). Illustration by Joe Tomelleri. Upper Snake River, introduced in Owyhee River.

SUCKERS
family Catostomidae

Suckers are widespread; larvae are numerous in shallow edges of rivers and lakes in late spring and early summer, and juveniles and adults are important prey of many piscivorous birds such as osprey. Four species in Oregon are on the Endangered Species List.

Suckers have large papillose lips and a longer trunk than minnows. The overall body form is superficially similar among many species, and care must be taken to ensure proper identification. Males have a somewhat pointed anal fin, while females have a more rounded anal fin.

Top, Shortnose Sucker, 15 mm, A09-149; *middle*, Shortnose Sucker, 75 mm, OS 18418; *bottom*, Klamath Smallscale Sucker, male, 211 mm, OS 18409.

KEY TO CATOSTOMIDAE

1a. Lower lips thin, with relatively few papillae and with a noticeable gap separating the 2 lobes; spawners with dark black lateral stripe 2

1b. Lower lips moderate to thick, with numerous rows of papillae and with each lobe making partial or complete contact along the midline; spawners with red or black lateral stripe 3

2a. Body elongate, with a trunk tending toward a uniform height or rising to the dorsal base; snout long, angular; 44–48 post-Weberian vertebrae; in specimens > 200 mm, 23–37 gill rakers .... *Deltistes luxatus*

Lost River Sucker, *Deltistes luxatus* Cope, 1879. 442 mm, OS 12812. Illustration by Joe Tomelleri. Klamath Basin, mostly Upper Klamath Basin.

2b. Body moderately elongate, generally rising toward the dorsal; snout moderate; 41–44 post-Weberian vertebrae; in specimens > 200 mm, 30–45 gill rakers .... *Chasmistes brevirostris*

Shortnose Sucker, *Chasmistes brevirostris* Cope, 1879. *Above*, 75 mm, OS 18418; *below*, 365 mm, OS 5306. Illustration by Joe Tomelleri. Klamath Basin, mostly Upper Klamath Basin.

3a. Large pelvic axillary process; interradial membranes of caudal fin clear; lower lip with a shallow or no median notch, usually a deep break at the corners of the mouth separating the upper and lower lips (no break in Lahontan specimens)...................................... *Catostomus bondi*

Cascadian Sucker, *Catostomus bondi* Smith, Stewart and Carpenter, 2013. Male, 235 mm, Willamette River. Part of Mountain Sucker (*C. platyrhynchus*) complex, Columbia Basin.

3b. Pelvic axillary process absent or represented by a fold, which may be prominent; interradial membranes of caudal fin dark; lower lip with a shallow or deep median notch, no deep break at the corners of the mouth separating the upper and lower lips 4

4a. Lower lips joined over about half their length or less............... 5

4b. Lower lips separated or almost completely separated............... 7

5a. Scales very small, more than 90 in lateral line; more than 30 gill rakers on first arch.................. *Catostomus columbianus*

Bridgelip Sucker, *Catostomus columbianus* (Eigenmann & Eigenmann, 1893). Illustration by Joe Tomelleri. Columbia Basin upstream from Hood River.

5b. Scales small, fewer than 91 in lateral line; fewer than 30 gill rakers on first arch.. 6

6a. Dorsal rays 11–13; peritoneum light; Klamath/Rogue drainages .... *Catostomus rimiculus*

Klamath Smallscale Sucker, *Catostomus rimiculus* Gilbert & Snyder, 1898. Female, 270 mm, Spencer Creek, released. Rogue River and Lower Klamath Basin.

6b. Dorsal rays 8–11; peritoneum dark;
Lahontan drainages.................................. *Catostomus tahoensis*

Tahoe Sucker, *Catostomus tahoensis* Gill & Jordan, 1878. 150 mm, OS 18031, McDermitt Creek. Quinn River in Oregon.

7a. Lateral fields of scales with numerous slanting radii; spawners with red sides; Goose Lake/Sacramento drainages 8

7b. Lateral fields of scales with either no radii or extensions of anterior radii; spawners with red, black, or golden sides; Columbia and Klamath drainages.................................. 9

8a. Adults small, rarely exceeding 18 cm and maturing at 8–12 cm; dorsal rays usually 10–11; scales smaller, usually 80–92 in lateral line; vertebrae usually 40–42
.... *Catostomus microps*

Modoc Sucker, *Catostomus microps* Rutter, 1908. Male, 174 mm, OS 15855, Thomas Creek. Goose Lake Basin.

8b. Adults larger, up to 50 cm and maturing at 20 cm; dorsal rays usually 11–13; scales larger, usually < 76 in lateral line; vertebrae usually 43–45................*Catostomus occidentalis*

Sacramento Sucker, *Catostomus occidentalis* Ayres, 1854. Illustration by Joe Tomelleri. Goose Lake Basin.

9a. Dorsal rays 6–10, usually 9–10; mature males with bright white ventrum extending from snout to caudal peduncle, with a red band about 6 scales wide above the lateral line on a black or dark background .... *Catostomus warnerensis*

Warner Sucker, *Catostomus warnerensis* Snyder, 1908. Male, 280 mm, OS 6359. Illustration by Joe Tomelleri. Warner Basin and stocked in Summer Lake Basin.

9b. Dorsal rays 10–16; mature males with white to yellow ventrum 10

10a. Gill rakers 29–40, usually 31–38 in fish > 200 mm; Klamath Basin .... *Catostomus snyderi*

Klamath Largescale Sucker, *Catostomus snyderi* (Gilbert, 1898). 385 mm, OS 12811. Illustration by Joe Tomelleri. Klamath Basin, mostly Upper Klamath Basin.

10b. Gill rakers 24–32 in fish > 200 mm; Tyee and Columbia Basins 11

11a. Dorsal rays 10–14, usually 11–13; infraorbital pores 14–38; caudal peduncle depth moderate, 6.4%–9.3% SL (75% of specimens less than 7.3% SL) ....*Catostomus tsiltcoosensis*

Tyee Sucker, *Catostomus tsiltcoosensis* Evermann & Meek, 1898. Male, 395 mm, Woahink Creek. Coastal streams from Siuslaw River to Sixes River. The 4 major streams in its distribution appear to contain independent genetic lineages, with the Coquille River the most divergent.

11b. Dorsal rays 12–15, usually 13–14; infraorbital pores 33–50; caudal peduncle depth shallower, 5.1%–8.7% SL (75% of specimens greater than 7.7% SL) .... *Catostomus macrocheilus*

Largescale Sucker, *Catostomus macrocheilus* Girard, 1856. *Above*, juvenile, 33 mm; *below*, male, 420 mm, Willamette River. Columbia Basin and Nehalem River.

LOACHES
family Cobitidae

Two species of nonnative loaches have been collected in the state. The Oriental Weatherfish, *Misgurnus anguillicaudatus*, is established in the Clackamas, Willamette, and lower Columbia and several Snake River tributaries. A specimen tentatively identified as Finescale Loach, *M. mizolepis*, was collected in 1977 from an irrigation ditch in the Powder River drainage, but there are no recent records. These are likely aquarium hobby releases.

Oriental Weatherfish, *Misgurnus anguillicaudatus* (Cantor, 1842). 142 mm, OS 15473, Carter Ditch, Snake River.

PACUS
family Characidae

Two species of Pacus, Pirapatinga (*Piaractus brachypomus*) and Caranha (*P. mesopotamicus*), have been collected in Oregon, and a third species, Tambaqui (*Colossoma macropomum*), is questionable. The Pacus have been misidentified as piranha in some news reports, but one Red-Bellied Piranha (*Pygocentrus nattereri*) was caught in 2003 and a photo and story published in the *Oregonian* newspaper. These are likely aquarium hobby releases. All live captures of characids in Oregon have been in summer, and it is unlikely any could survive winters unless in natural warm springs or artificially heated water.

Pirapatinga, *Piaractus brachypomus* (Cuvier, 1818). 201 mm, OS 11491, Willamette River.

CATFISHES
family Ictaluridae

Catfishes and bullheads are distinctive nonnative fishes first introduced to Oregon in 1880 from the eastern United States. They are also present in the fossil record of Oregon. Some are valued as sport and food fish, but they may also have played a role in the decline of some native fishes.

KEY TO ICTALURIDAE

1a. Adipose fin continuous with caudal fin; adults small .... *Noturus gyrinus*

Tadpole Madtom, *Noturus gyrinus* (Mitchill, 1817). Illustration by Joe Tomelleri. Introduced to Snake River.

1b. Adipose fin distinctly separate from caudal fin; adults moderate to large 2

2a. Caudal fin deeply forked... 3

2b. Caudal fin rounded, slightly to moderately forked 4

3a. Anal fin rays usually 30–36; anal fin margin straight; no spots on body.. *Ictalurus furcatus*

Blue Catfish, *Ictalurus furcatus* Valenciennes in Cuvier & Valenciennes, 1840. Illustration by Joe Tomelleri. Introduced in Snake and Willamette Rivers.

3b. Anal fin rays usually 25–30; anal fin margin rounded; body with spots in juveniles and small adults .... *Ictalurus punctatus*

Channel Catfish, *Ictalurus punctatus* (Rafinesque, 1818). Illustration by Joe Tomelleri. Introduced in Columbia Basin, more abundant in eastern Oregon.

4a. Caudal fin moderately forked; hard ridge (supraoccipital bone) from head does not reach base of first dorsal fin and creates a soft spot in front of fin................. *Ameiurus catus*

White Catfish, *Ameiurus catus* (Linnaeus, 1758). Illustration by Joe Tomelleri. Introduced in Columbia Basin, scattered.

4b. Caudal fin rounded or slightly forked 5

5a. Head broad and flat; lower jaw projecting beyond upper; upper caudal lobe pale; coloring uniformly dark to mottled; large fish, adults to 90 cm TL............ *Pylodictis olivaris*

Flathead Catfish, *Pylodictis olivaris* (Rafinesque, 1818). Illustration by Joe Tomelleri. Introduced in Snake River.

5b. Head rounded, rarely flat; lower jaw not projecting beyond upper; upper and lower caudal lobes similarly colored; moderate-sized fish, < 50 cm TL................................. 6

6a. Anal rays 24–27; chin barbels pale; body and fins uniformly dark, belly white with yellow fringe .... *Ameiurus natalis*

Yellow Bullhead, *Ameiurus natalis* (Lesueur, 1819). 134 mm, OS 18435, Willamette River. Introduced, widespread.

6b. Anal rays 20–24; chin barbels dark 7

7a. Gill rakers 13–15; caudal fin base uniformly colored; body and fins uniformly dark .... *Ameiurus nebulosus*

Brown Bullhead, *Ameiurus nebulosus* (Lesueur, 1819), 153 mm, OS 18436, Willamette River. Introduced, widespread.

7b. Gill rakers 16–19; caudal base often with rectangular pale area; body and fins uniformly dark .... *Ameiurus melas*

Black Bullhead, *Ameiurus melas* (Rafinesque, 1820). Illustration by Joe Tomelleri. Introduced in eastern Oregon, distribution elsewhere uncertain, as it is often confused with Brown Bullhead.

PIKES
family Esocidae

Three pikes, the Northern Pike (*Esox lucius*), the Redfin or Grass Pickerel (*E. americanus*), and the Tiger Muskellunge, a hybrid of the Northern Pike and Muskellunge (*E. masquinongy*), have been introduced in Oregon.

The Northern Pike has rows of pale spots on the sides, spots or streaks on the median fins, 10–13 pelvic rays, and no dark subocular bar. In upper Columbia River from introduction in Idaho.

The Redfin Pickerel has vertical to oblique banding on the sides, red fins with no spots or streaks, 8–10 pelvic rays, and a dark subocular bar. In upper Columbia River from introduction in Washington.

The Tiger Muskellunge (not shown) is a hybrid of *E. lucius* and Muskellunge (*E. masquinongy*) and has dark vertical to oblique bands like redfin pickerel but more rounded fins, and only the upper half of the opercle is scaled. They were stocked in Phillips Reservoir in the Powder River Basin, and there are reports of collections below Willamette Falls.

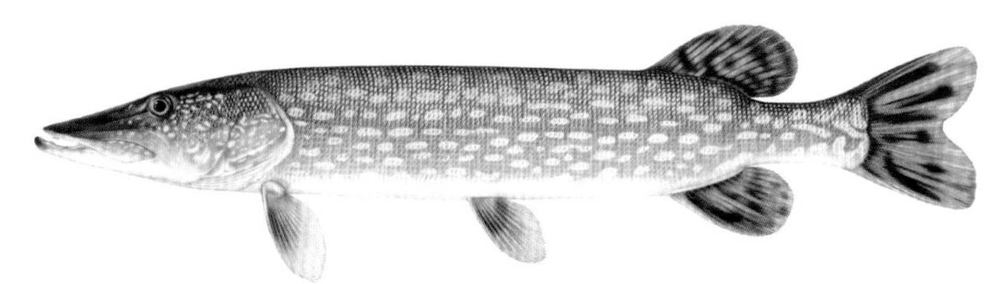

Northern Pike, *Esox lucius* Linnaeus, 1758. Illustration by Joe Tomelleri.

Redfin Pickerel, *Esox americanus* Gmelin, 1789. Illustration by Joe Tomelleri.

SMELTS
family Osmeridae

Smelts are mostly marine, but two anadromous smelts spawn in freshwater in Oregon. The Longfin Smelt (*Spirinchus thaleichthys*) has an incomplete lateral line extending to the dorsal fin origin, eye diameter about equal to snout length, and a large pectoral fin extending almost to the pelvic fins.

The Eulachon or Candlefish (*Thaleichthys pacificus*) has a complete lateral line, eye diameter smaller than snout length, and shorter pectoral fin extending about half the distance to the pelvic fins. Large runs historically came into the Klamath, lower Columbia, and Sandy Rivers, and Native Americans valued them for their fat content. US populations are now listed as threatened. Although there are many stories about the origin of the word "Oregon," some believe it came from the Native American word *ooligan*, referring to the highly prized grease made from Eulachon.

Longfin Smelt, *Spirinchus thaleichthys* (Ayres, 1860). 80 mm SL, OS 13679, off Oregon coast.

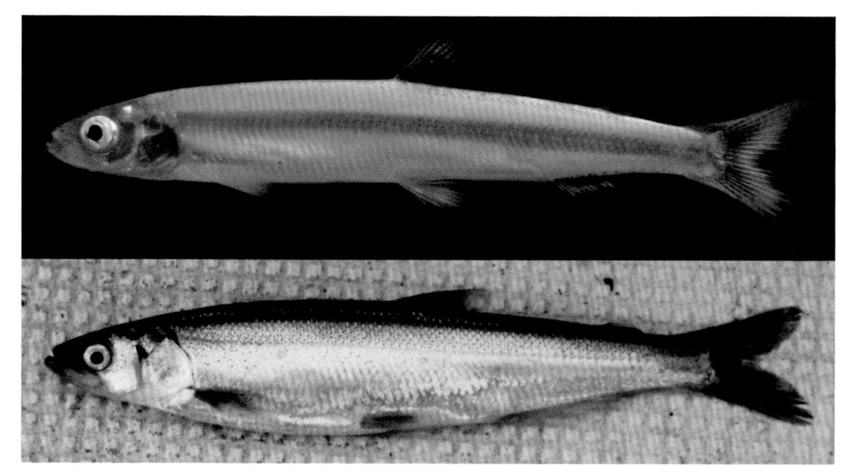

Eulachon, *Thaleichthys pacificus* (Richardson, 1836). *Above*, juvenile, Salmon River; *below*, adult, ca. 120 mm, offshore, Oregon.

SALMON AND TROUTS
family Salmonidae

Salmonids, including trout, salmon, char, and whitefish, are the iconic fish of the Pacific Northwest and Oregon and are found in most regions of the state, even the dry Alvord Basin. The biodiversity of salmonids has a complex nomenclature related to the diversity of their life-history patterns. There are terms for different forms of the same sex (precocious, jack, adult), different juvenile residence times in freshwater (stream-type, ocean-type), different tolerance to saltwater (anadromous, resident), different migration timing, different spawning strategies, and numerous local variants. On top of this diversity, humans have added a hodgepodge of hatchery stocking and movement of fish between basins for well over 100 years. Two introduced forms not included herein, Arctic Charr and hybrid Tiger Trout, may be encountered.

Species names are generally settled, but the complexes of rainbow trouts and cutthroat trouts are often subdivided into subspecies, and even "minor subspecies." Many of these are in drainages that have not

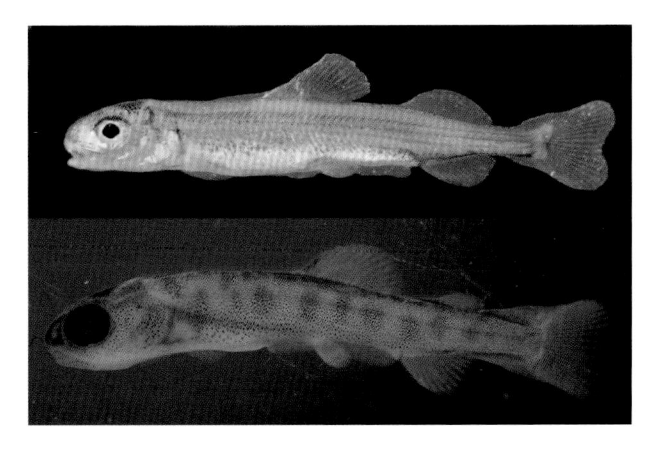

Juvenile salmonids. *Above*, Mountain Whitefish, 20.5 mm, Willamette River; *below*, Rainbow Trout, 18.5 mm, Rogue River.

been connected for millions of years. Within large drainages, such as the Klamath or Columbia, there can be 2 or more subspecies. Adding to the confusion, both trouts can readily hybridize across species.

Early stages can be difficult to distinguish among the trouts, but subfamilies are distinctive. Pollard et al. (1997) give an identification aid for coastal species that performs poorly on interior forms.

KEY TO SALMONIDAE

1a. Scales large, about 70–100 in lateral line; snout prominent; mouth inferior or subterminal, adults without teeth on the jaws 2

1b. Scales small, usually > 110 in lateral line; mouth terminal, with well- developed teeth in jaws 3

2a. Snout pointed, "pinched"; single flap of skin between nostrils; adipose fin large, length of its base equal to or greater than caudal peduncle depth, the base becoming proportionally smaller with growth; gill rakers short, 20–25; mouth small, upper jaw not reaching orbit .... *Prosopium williamsoni*

Mountain Whitefish, *Prosopium williamsoni* (Girard, 1856). *Above*, juvenile, 49 mm, OS 18447; *below*, adult, 114 mm, OS 18431, Willamette River. Columbia Basin. Recent genetic work suggests this very wide-ranging fish requires more taxonomic research, especially regarding the related Pygmy Whitefish.

2b. Snout broad, overhanging mouth; 2 small flaps of skin between nostrils; adipose fin smaller; gill rakers long, 23–33; mouth larger, upper jaw about reaching anterior of orbit...*Coregonus clupeaformis*

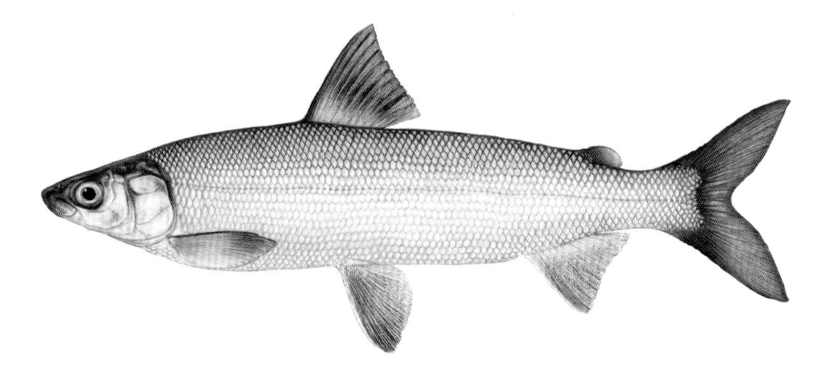

Lake Whitefish, *Coregonus clupeaformis* (Mitchill, 1818). Introduced. Illustration by Joe Tomelleri. Upper Columbia River above McNary Dam.

3a. Anal fin with 13–19 rays; anal fin base longer than dorsal fin base; parr without pigmented first dorsal ray........... 4

3b. Anal fin with 8–12 rays; anal fin base shorter than dorsal fin base; first dorsal ray of parr black or lightly pigmented.. 8

4a. Distinct black spots on back and caudal fin 5

4b. No black spots on back or caudal fin 7

5a. Spots oblong, relatively large; young without parr marks ....*Oncorhynchus gorbuscha*

Pink Salmon, *Oncorhynchus gorbuscha* (Walbaum, 1792). Male. Illustration by Joe Tomelleri. Coastal streams as strays.

5b. Spots small, irregular; young with parr marks.......................... 6

6a. Caudal spots only on upper lobe of caudal fin; white gums at base of teeth in lower jaw; parr marks long and narrow, their width less than space between them; parr with falcate anal fin with a white leading edge followed by black; adipose fin in parr with dark edge and pigmented center .... *Oncorhynchus kisutch*

Coho Salmon, *Oncorhynchus kisutch* (Walbaum, 1792). *Above*, juvenile, photo credit, Bruce Miller; *below*, adult male, illustration by Joe Tomelleri. Coastal rivers.

6b. Caudal spots on upper and lower lobes of caudal fin; black gums at base of teeth in lower jaw; parr marks more oval, their width equal to or wider than space between them; parr without falcate anal fin but with a white leading edge that is not followed by black; adipose fin in parr with dark edge and clear center ....*Oncorhynchus tshawytscha*

Chinook Salmon, *Oncorhynchus tshawytscha* (Walbaum, 1792). *Above*, juvenile, Salmon River; *below*, adult male. Illustration by Joe Tomelleri. Coastal rivers, Columbia and Willamette Rivers.

7a. Gill rakers short and smooth, 16–26; spawners with black head and irregular purple blotches on sides; parr with parr marks mostly above lateral line and sides green .... *Oncorhynchus keta*

Chum Salmon, *Oncorhynchus keta* (Walbaum, 1792). Male. Illustration by Joe Tomelleri. Coastal rivers, especially Tillamook Bay, Columbia River.

7b. Gill rakers long, serrated, and more numerous, 29–44; spawners with red sides and green head; parr with some parr marks bisected by lateral line and sides white .... *Oncorhynchus nerka*

Sockeye Salmon, *Oncorhynchus nerka* (Walbaum, 1792), landlocked forms known as Kokanee. Male. Illustration by Joe Tomelleri. Columbia, Willamette, Deschutes Rivers.

8a. Spots on back and sides dark on a light background................ 9

8b. Spots on back and sides light on a dark background.............. 17

9a. Irregular red spots along lateral line, pale halos around some spots on sides; adipose fin in parr orange to red .... *Salmo trutta*

Brown Trout, *Salmo trutta* Linnaeus 1758. *Top*, parr, 20 mm, Rogue River; *middle*, subadult, 129 mm, Crooked Creek, OS 18808; *bottom*, adult, ca. 400 mm, Crooked Creek, released. Introduced, especially in Cascade Lakes.

9b. No red spots along lateral line, no pale halos around spots on sides; adipose fin in parr clear or with dark border 10

10a. Origin of dorsal fin distinctly in advance of middle of body; x-shaped spots on sides; in parr adipose fin clear and pectoral fin long, extending to vertical from origin of dorsal fin .... *Salmo salar*

Atlantic Salmon, *Salmo salar* Linnaeus, 1758. Illustration of landlocked form by Joe Tomelleri. Introduced in Hosmer Lake.

10b. Origin of dorsal fin in middle of body; spots on sides irregular to round, seldom x-shaped; in parr adipose fin with dark border and pectoral fin shorter, not extending to vertical from origin of dorsal fin 11

11a. Distinctive red/orange slash under lower jaw; basibranchial teeth usually present behind tongue, upper jaw extending well beyond orbit 12

11b. Usually no red/orange slash under lower jaw; basibranchial teeth usually absent; upper jaw reaching posterior margin of orbit, sometimes extending past orbit in anadromous and eastern Oregon forms 15

12a. Gill rakers 20–28; usually 2 epurals in caudal skeleton in Oregon; spots generally rounded, moderate to large; parr marks usually more narrow than rounded; Lahontan Basin .... *Oncorhynchus henshawi*

Lahontan Cutthroat Trout, *Oncorhynchus henshawi* (Gill & Jordan in Jordan, 1878). Illustration of lake form by Joe Tomelleri. In Quinn River, introduced in Alvord Basin. Usually considered a subspecies, *O. clarkii henshawi.*

12b. Gill rakers 15–21; usually 3 epurals in caudal skeleton; spots irregularly shaped, small to moderate; parr marks usually more rounded than narrow; coastal and Columbia basins 13

13a. Adults profusely covered with small spots; anal fin tipped with white; coastal and Columbia basins .... *Oncorhynchus clarkii*

Coastal Cutthroat Trout, *Oncorhynchus clarkii* (Richardson, 1837). *Above*, juvenile, 84 mm, OS 18403, Clatskanie River; *below*, adult, ca. 180 mm, Fall Creek. Widespread in coastal streams. Usually considered a subspecies, *O. clarkii clarkii.*

13b. Adult spotting sparse on anterior part of body; posterior spots larger; anal fin without white tip.................................. 14

14a. Arc-shaped area from anal to pectoral with few or no spots; John Day Basin isolate .... *Oncorhynchus lewisi*

Westslope Cutthroat Trout, *Oncorhynchus lewisi* (Girard, 1856). Illustration of a male by Joe Tomelleri. Widespread in northern Rocky Mountains; only in John Day River in Oregon. Usually considered a subspecies, *O. clarkii lewisi.*

14b. Few spots on body in front of anus ... *Oncorhynchus alvordensis*

Alvord Cutthroat Trout, *Oncorhynchus alvordensis* Sigler and Sigler, 1987. Illustration by Joe Tomelleri. Considered extinct, taxonomic status uncertain, and usually considered a subspecies, *O. clarkii alvordensis*. Possibly introduced in Guano Creek with 1 or more other forms of the Cutthroat Trout complex.

15a. Lower side bright gold; belly, lateral band, and branchiostegals orange red................ *Oncorhynchus aguabonita*

Golden Trout, *Oncorhynchus aguabonita* (Jordan, 1893). Illustration by Joe Tomelleri. Introduced, scattered, especially in lakes in Wallowa County.

15b. Coloration highly variable, not as above 16

16a. Parr marks usually more rounded than elliptical; interior basins .... *Oncorhynchus mykiss newberryi*

Redband Trout, *Oncorhynchus mykiss newberryi* (Girard, 1859). *Top*, 25 mm, OS 18806, Williamson River; *middle*, 51 mm, OS 18408, Spencer Creek; *bottom*, 95 mm, OS 18806, Williamson River. Interior basins. More work on morphological variability is needed.

16b. Parr marks usually more elliptical than rounded; coastal and Columbia basins......................*Oncorhynchus mykiss mykiss*

Rainbow Trout/Steelhead, *Oncorhynchus mykiss mykiss* (Walbaum, 1792). *Above*, 39 mm, OS 17969, Coquille River; *below*, 150 mm, OS 18404, Clatskanie River. Coastal streams.

17a. Wavy, wormlike lines on dorsal fin and along back; blue halos around spots on lower side; pelvic and anal fins with wide immaculate white anterior margin, followed by black edge and orange.............................*Salvelinus fontinalis*

Brook Trout, *Salvelinus fontinalis* (Mitchill, 1814). *Top*, 16 mm, Rogue River; *middle*, 65 mm, OS 18803, Miller Creek; *bottom*, about 200 mm, Wood River. Introduced, widely distributed in cold streams.

17b. Dorsal fin dusky, sometimes with spotting but without wavy lines; wavy, wormlike lines present or absent along back; spots without blue halos; pelvic and anal fins with narrow whitish anterior margin, with little or no black edge, followed by orange 18

18a. Tail deeply forked; irregular light spots on sides, dorsal fin, and caudal fin; wavy, wormlike lines present along back .... *Salvelinus namaycush*

Lake Trout, *Salvelinus namaycush* (Walbaum, 1792). Illustration of a Lake Superior specimen by Joe Tomelleri. Introduced, especially in Cascade Lakes.

18b. Tail not deeply forked; regular colored spots on sides of adults; dorsal fin and caudal fin without spots (hybrids with *S. fontinalis* may have spotted dorsal fins) .... *Salvelinus confluentus*

Bull Trout, *Salvelinus confluentus* (Suckley, 1858). *Above*, 69 mm, hatchery raised; *below*, 156 mm, Three-Mile Creek, Klamath Basin, released. A sister species, Dolly Varden, *S. malma* (Walbaum, 1792), is often confused with Bull Trout but does not extend farther south than northern Washington. Klamath and Columbia Basins.

SAND ROLLER
family Percopsidae

The Sand Roller (*Percopsis transmontana*) is a Columbia River endemic. In Oregon its distribution is restricted to the Columbia River main stem and the Willamette River and many of its tributaries. It is active at night and may seek deeper water during the day; consequently, it is not encountered in proportion to its abundance.

Sand Roller, *Percopsis transmontana* (Eigenmann & Eigenmann 1892). 62.5 mm SL, OS 17965, Willamette River. Columbia Basin, Willamette River.

FRESHWATER CODS
family Lotidae

The Burbot is a freshwater member of one of the cod families. It is most common in the upper Columbia River; only a single museum specimen has ever been collected from Oregon, but there are recent reports of a few specimens in the Columbia River in Oregon.

Burbot, *Lota lota* (Linnaeus, 1758). Illustration by Joe Tomelleri.

SILVERSIDES
family Atherinopsidae

Two species of silversides are included. The Topsmelt (*Atherinops affinis*) can be abundant in estuaries and lower parts of coastal rivers but does not spawn in freshwaters of Oregon.

The nonnative Mississippi Silverside (*Menidia audens*) was introduced to California in the late 1980s (sometimes misidentified as Inland Silverside, *M. beryllina*). A single specimen was collected by Keith Marine in a screw trap in the Link River, Klamath County, in 2008. In California reservoirs, Mississippi Silversides have replaced small nearshore fishes, especially juveniles, and have had negative effects on aquatic ecosystems. Peter Moyle (University of California–Davis) predicted that they would probably spread from introductions by bait fishermen. They are a potential invader that should be monitored.

Topsmelt, *Atherinops affinis* (Ayres, 1860). 176 mm, OS 7689, Yaquina Bay.

Mississippi Silverside, *Menidia audens* Hay, 1882. Photo credit, René Reyes.

KILLIFISHES
family Fundulidae

There are 2 nonnative species of killifishes. The Banded Killifish (*Fundulus diaphanus*) is locally common in off-channel areas of the Willamette and lower Columbia Rivers and in some tributaries of the Snake River and could have been introduced either as an aquarium release or for mosquito control.

The Rainwater Killifish (*Lucania parva*) is restricted to the Yaquina River and appears to have been introduced with oysters or oyster spat from the eastern United States.

Banded Killifish, *Fundulus diaphanus* (Lesueur, 1817). Burnt River, August 2005.

Rainwater Killifish, *Lucania parva* (Baird & Girard, 1855). Male, ca. 30 mm, Neuse River, North Carolina. Photo credit, Jesse Bissette.

LIVE-BEARERS
family Poeciliidae

Mosquitofish, native to central and eastern North America, have been introduced worldwide as a vector control agent. There are 2 species, Western Mosquitofish (*Gambusia affinis*), usually with 6 dorsal fin rays and 9 anal fin rays, and Eastern Mosquitofish (*G. holbrooki*),

Western Mosquitofish, *Gambusia affinis* (Baird & Girard, 1853). Top, juvenile, 7 mm; middle, male, 21 mm; bottom, female, 38 mm, (both adults OS 18446).

usually with 7 dorsal fin rays and 10 anal fin rays. Most Oregon specimens are Western Mosquitofish, but the 2 species hybridize readily and the source(s) of introductions are unknown. It is therefore possible that both species and their hybrids could be found.

Females are about twice the size of males, to about 50 mm SL, with males to about 25 mm SL. Fertilization is internal and facilitated by the male's modified anal fin (gonopodium), whose structure also differs between these species (*G. affinis* lacks teeth on the third gonopodial ray). After mating, females retain live sperm that can fertilize several batches of eggs. Gestation is 3–4 weeks, and a single female can produce several broods of 30–70 young each in a summer. Young can mature as quickly as 4–6 weeks, and early broods can reproduce in the summer of their birth.

Introduced mosquitofish have been implicated in the declines of native fishes, amphibians, and fairy shrimp. The mechanism is thought to be predation, often on egg or larval stages. Another live-bearer, the Guppy, *Poecilia reticulata*, is a popular aquarium fish and might be expected in state waters, but we have no documented records.

STICKLEBACKS
family Gasterosteidae

The Threespine Stickleback is a coastal-oriented fish found in the Pacific and Atlantic basins of the Northern Hemisphere. It can be marine, anadromous, stream-resident, or lacustrine. Isolated freshwater populations can diverge from the anadromous form and rapidly become reproductively isolated. The resulting species are usually not formally recognized, producing a species complex in which all members are given the same scientific name, *Gasterosteus aculeatus*.

Threespine Sticklebacks are one of the best-known small fishes in North America and have been the subject of numerous scientific studies, especially of their spawning behavior. There is great diversity in their behavior, but typically, males build nests on bottom substrate that they cover with plant material glued together with a substance from their kidneys. Males are territorial and develop nuptial colors, usually red throats and blue eyes, and entice females to spawn in their nests. After fertilization, males guard the eggs and fan them with their pectoral fins. Individual life spans are usually 1–5 years.

Threespine Stickleback, *Gasterosteus aculeatus* Linnaeus, 1758. *Above*, mature male, 47 mm; *below*, mature female, 43 mm, both OS 17985, Elk Creek, Umpqua River.

SCULPINS
family Cottidae

Sculpins are a family of mostly marine fishes, but there is a freshwater genus, *Cottus*, in North America and Eurasia. Two species of *Cottus* are at least partly dependent on ocean or estuarine habitats; the young of *C. aleuticus* are reared and some adult *C. asper* spawn in estuaries or near shore. Two marine sculpins, Sharpnose Sculpin and Staghorn Sculpin, may be found in estuaries but do not spawn in freshwater and are not included here.

Most *Cottus* species are adapted to a bottom-oriented life and have similar shapes, with a large, flat head and mouth; eyes high on the head, a margin of iris around the pupil that may be red in life; expanded fanlike pectoral fins; a robust, naked or prickled body; a long anal fin and second dorsal fin; and a rounded or truncate caudal fin. The Klamath endemic *C. princeps* is adapted for more open water and is the most distinct, with a slender body and upturned mouth. Western North American sculpins are notoriously difficult to identify and their taxonomy is unsettled. Local differences, ontogenetic changes in characters, hybridization, and currently unrecognized species contribute to the confusion. Although the attached key should be helpful, it is advisable to examine multiple specimens from a drainage, and to save voucher specimens that can be compared to museum specimens.

Sculpins are diverse and ecologically important in Oregon, but lack of confidence in identification has precluded gaining more knowledge. They may be the dominant vertebrate in many streams. They prey on many other fishes and may have a detrimental impact on salmonids under some conditions. They are also important prey for many larger fish, including salmonids, as well as for other animals, and may be sentinels of water quality.

Cottus rhotheus, 45 mm, in spawning coloration, OS 18427, Willamettte River.

KEY TO COTTIDAE

Sculpins are difficult to identify, and this key is not very efficient, especially the first couplet. Some changes with size will make this couplet lead to erroneous identifications, and some specimens will be intermediate. If a specimen does not seem to match the final identification, try the alternate choice for this couplet. Comparisons with museum specimens and familiarity with the species in a given drainage are often necessary to increase confidence in sculpin identification.

1a. Mouth large, with opercles compressed and not flared, the posterior tips of the maxillaries behind a vertical through the anterior margin of the lens and usually under the posterior ¾ of the eye 2

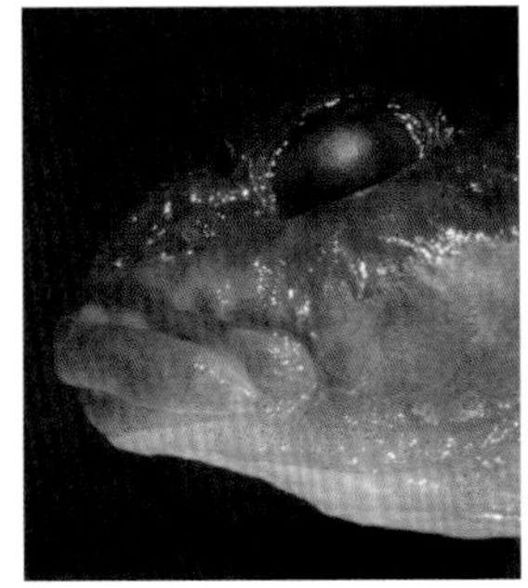

1b. Mouth small, with opercles compressed and not flared, the posterior tips of the maxillaries in front of a vertical through the anterior margin of the lens and usually under the anterior ¼ of the eye 9

2a. Anal fin long, 15–19 rays (rarely 15–16); second dorsal rays 19–23; body well covered with prickles, especially in specimens from inland waters and in young individuals from coastal waters .... *Cottus asper*

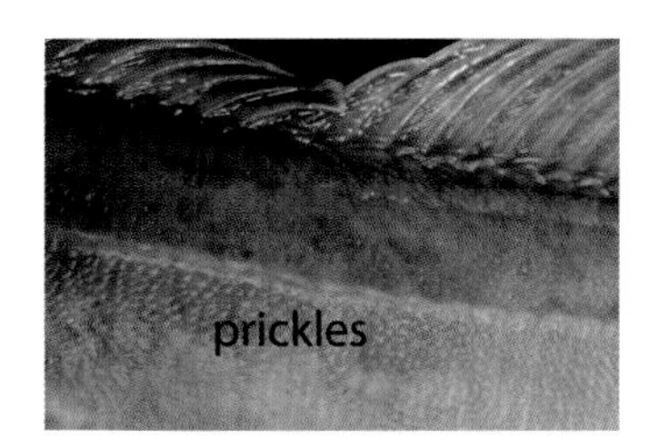

Prickly Sculpin, *Cottus asper* Richardson, 1837. 92 mm, OS 5933, Whiskey Creek. Coastal streams, Willamette River, and Columbia below Snake River.

2b. Anal fin shorter, rays 11–17; second dorsal rays 15–20 (rarely 20); body naked to well covered with prickles 3

3a. Body rather uniformly slender with little front-to-rear taper; a prickle patch on side under pectoral fin, lateral line complete to under posterior part of dorsal fin (about 15th ray), then interrupted or underdeveloped; head distinctly small, usually 25%–30% SL; body light brown with indistinct saddles under second dorsal....... *Cottus confusus*

Shorthead Sculpin, *Cottus confusus* Bailey & Bond, 1963. 57 mm, OS 15593, Plympton Creek. Scattered, lower Columbia and upper sections of Columbia tributaries.

3b. Body usually more robust anteriorly; lateral line complete or incomplete; head larger, usually 30%–35% SL or larger; body coloration variable, from mottled to having a strong saddle pigmentation 4

4a. Two prominent, forward-slanting saddles under second dorsal usually extending below lateral line; 2 fainter saddles joined under first dorsal; 2 chin pores; top of head and eyes heavily nubbled; lateral line complete; caudal truncate to subtruncate .... *Cottus rhotheus*

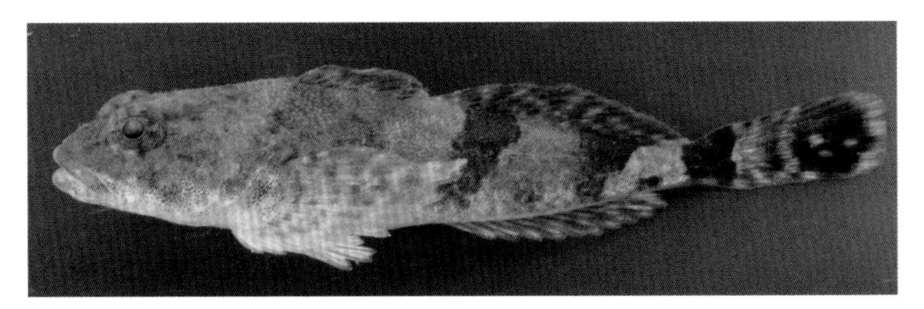

Torrent Sculpin, *Cottus rhotheus* (Smith, 1882). 42 mm, OS 18427, Willamette River. Columbia Basin and coastal streams south to Trask River.

4b. Saddles under second dorsal present or indistinct, not slanting forward and not extending below lateral line; usually 1 or 2 chin pores; lateral line variable, sometimes complete 5

5a. Pelvic rays usually 3 (75% of specimens), if 4, innermost ray reduced; usually 1 chin pore; anal rays 15–17 .... *Cottus marginatus*

Margined Sculpin, *Cottus marginatus* (Bean, 1881). 52 mm, OS 9469, Walla Walla River. Columbia Basin, Umatilla River to Walla Walla River.

5b. Pelvic rays usually 4 6

6a. Dorsal fins slightly to well joined, connecting membrane up to ½ height of first ray in second dorsal; saddles under second dorsal fin indistinct; palatine teeth present .... *Cottus gulosus*

Riffle Sculpin, *Cottus gulosus* (Girard, 1854). 80 mm, OS 10534, Fogarty Creek. Coastal streams, lower Columbia and Willamette Rivers. Can be difficult to distinguish from *C. perplexus*, which has a smaller mouth and no palatine teeth. More research is needed; this is probably a different species than *C. gulosus* in the Sacramento River.

6b. Dorsal fins separate or barely joined; saddles under second dorsal fin distinct or not; palatine teeth present or absent 7

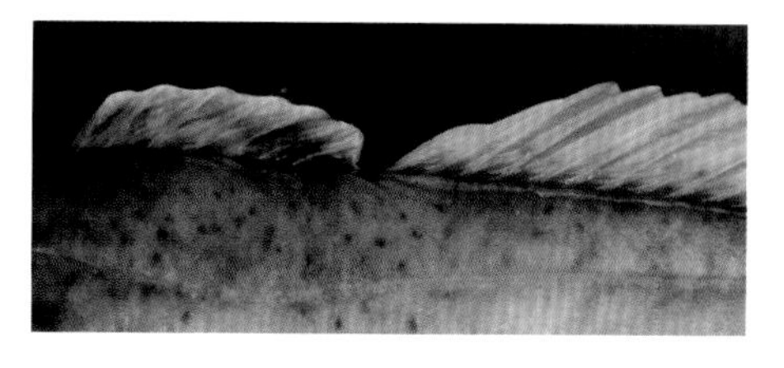

7a. Saddles under second dorsal fin indistinct; palatine teeth absent .... *Cottus pitensis*

Pit Sculpin, *Cottus pitensis* Bailey & Bond, 1963. 89 mm, OS 7000, Drews Creek. Goose Lake drainage.

7b. Saddles under second dorsal fin distinct; palatine teeth present .... Mottled Sculpin Complex (8a and b below)

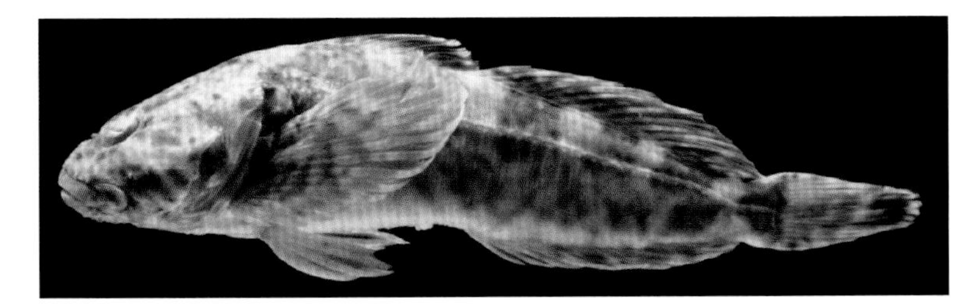

Mottled Sculpin complex, *Cottus hubbsi*. 70 mm, OS 0619, Wallowa River. More research is need on this complex; the following couplet separates 2 of the forms, which are sometimes considered subspecies, but there may be more than 2.

8a. Number of prickles on left side of body fewer than 30; lateral line about 90% complete, head usually smaller than predicted from the relationship HL = 0.2238 + 0.3115 * SL .... *Cottus bendirei*

Malheur Mottled Sculpin, *Cottus bendirei* (Bean, 1881). 82 mm, OS 16706, Poison Creek. Harney Basin and perhaps scattered localities in Malheur River and lower Columbia River.

8b. Number of prickles on left side of body more than 90; lateral line about 98% complete, head usually larger than predicted from the relationship HL = 0.2238 + 0.3115 * SL .... *Cottus hubbsi*

Columbia Mottled Sculpin, *Cottus hubbsi* Bailey & Dimick, 1949. 74 mm, OS 2546, Malheur River. Columbia and Harney Basins.

9a. Head sensory pores and mandibular pores very large, preorbital pores longer than interspaces; mandibular pores appearing more oval than circular; mouth directed upward; second dorsal fin rays 20–23; anal rays 16–18; prickles well developed; preopercular spine single and blunt or reduced; no palatine teeth; body slender; Upper Klamath Lake endemic .... *Cottus princeps*

Klamath Lake Sculpin, *Cottus princeps* Gilbert, 1898. *Above*, 27 mm, A09320; *below*, 76 mm, OS 14250, both Upper Klamath Lake. Upper Klamath Lake and nearby tributaries.

9b. Sensory pores on head not enlarged, mouth not directed upward 10

10a. One chin pore or 2 contiguous or fused pores (highly variable in *C. perplexus*) 11

10b. Two separate chin pores 12

11a. Posterior nostril tubular, equaling anterior in height; pale spot on top of caudal peduncle; first dorsal fin variable, clear, barred or with 2 large spots; preopercular spine single or with a second, shorter spine; lateral line complete; caudal peduncle long, about 2/3 length of anal fin base; spawners with orange spots with white margin located on dorsum at origin and insertion of second dorsal fin and at base of pectoral fin .... *Cottus aleuticus*

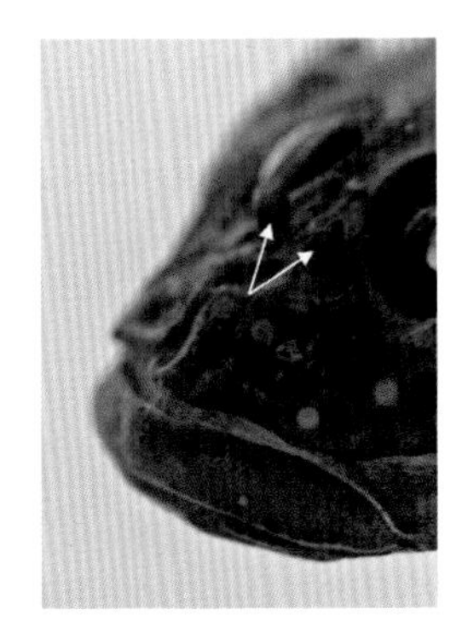

Coastrange Sculpin, *Cottus aleuticus* Gilbert, 1896. *Above*, live spawning colors, ca. 75 mm, Siletz River; *below*, 89 mm, OS 13877, Siuslaw River. Coastal streams and lower Columbia but not Willamette River.

11b. Prominent black spot on posterior of first dorsal fin (individuals in upper Willamette tributaries have a solid bar and larger mouth); posterior nostril short, not equaling anterior in height; some orange striping on margin of first dorsal fin and in membranes of second dorsal fin, but not on body; no palatine teeth................................ *Cottus perplexus*

Reticulate Sculpin, *Cottus perplexus* Gilbert & Evermann, 1894. 42 mm, OS 17986, Umpqua River. Coastal streams, lower Columbia and Willamette Rivers. Questionable records in eastern Oregon.

12a. Pelvic rays 3; body rather uniformly slender with little front-to-rear taper; lateral line complete to base of last dorsal fin ray; head distinctly small, especially in width; anal rays 14–17; usually 5–6 first dorsal spines; Upper Klamath Basin endemic... *Cottus tenuis*

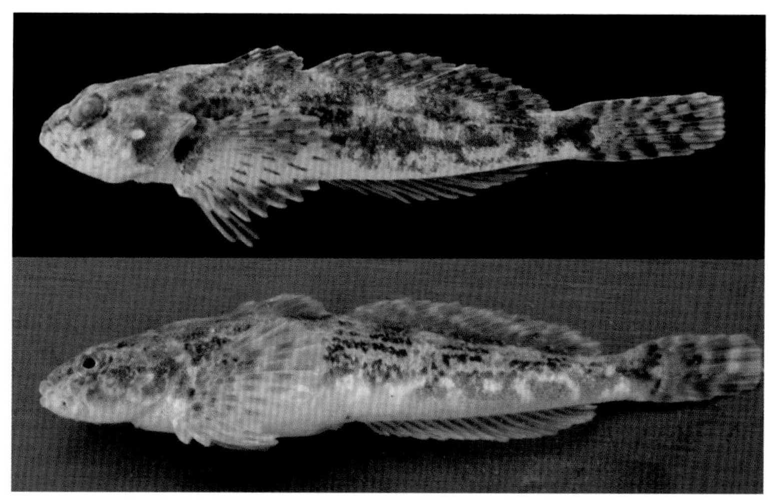

Slender Sculpin, *Cottus tenuis* (Evermann & Meek, 1898). *Above*, 35 mm, A09352, Upper Klamath Lake; *below*, 65 mm, OS 18809, Crooked Creek. Upper Klamath Lake and nearby tributaries.

12b. Pelvic rays usually 4, if 3 then anal rays 14 or fewer; head larger, more robust; body usually more robust anteriorly; lateral line complete or incomplete; usually 7–8 first dorsal spines 13

13a. Dorsal fins separate or slightly joined; first dorsal with anterior and posterior dark blotches; body slender; prickles absent in axil of pectoral fin; 2 chin pores; pale yellow spot at posterior base of second dorsal fin on caudal peduncle, saddle pigmentation poorly defined .... *Cottus beldingii*

Paiute Sculpin, *Cottus beldingii* Eigenmann & Eigenmann, 1891. 71 mm, OS 17094, Willamette River. Columbia and Snake Rivers and tributaries.

13b. Dorsal fins broadly joined; first dorsal fin with posterior dark blotch and uniform dark pigmentation on anterior membranes; body robust; prickles present or absent in axil of pectoral fin; Klamath Basin endemics 14

14a. Lateral line with 15–22 pores, stopping under anterior third of second dorsal fin; all axillary prickles below lateral line; upstream of Lake Ewauna .... *Cottus klamathensis*

Klamath Marbled Sculpin, *Cottus klamathensis* Gilbert, 1898. *Above*, 13 mm, A07231, Upper Klamath Lake; *below*, 90 mm, OS 18805, Williamson River. Upper Klamath Lake Basin.

14b. Lateral line with 22–29 pores, stopping under posterior third of second dorsal fin; a few axillary prickles above lateral line; downstream of Lake Ewauna, Klamath River .... *Cottus polyporus*

Siskiyou Marbled Sculpin, *Cottus polyporus* Daniels & Moyle, 1984. 86 mm, OS 7940, Klamath River. Often considered a subspecies of *C. klamathensis*, but both forms may co-occur in Lake Ewauna. Lower Klamath Basin.

SEA BASSES
family Moronidae

The nonnative Striped Bass, *Morone saxatilis*, an eastern North American food and game fish, was introduced into San Francisco Bay in 1879. It expanded its range and in 1914 was present in Coos Bay. It can now be found in all of the larger coastal rivers of Oregon, including the Columbia River.

Striped Bass, *Morone saxatilis* (Walbaum, 1792). Illustration by Joe Tomelleri.

SUNFISHES
family Centrarchidae

The nonnative basses and sunfishes were first introduced into Oregon in the 1800s. More recently, additional introductions have been made by state fisheries programs and by unauthorized "sportsmen."

Largemouth Bass. *Top*, 11 mm, A07-439, Upper Klamath Lake; *middle*, 21 mm, Calapooia River; *bottom*, 150 mm, OS 18812, Willamette River.

The Sacramento Perch is the only member of the family native to the West Coast, but it is also not a native Oregon species. The most recent addition is the Spotted Bass, which can easily be confused with the Largemouth Bass. At least 2 native fishes, the Oregon Chub and Umpqua Chub, as well as several amphibians, have been negatively impacted by centrarchids.

Male sunfishes can be brightly colored and build nests in shallow water where they are easily observed. They can hybridize readily.

KEY TO CENTRARCHIDAE

1a. Anal fin spines 5–8 2

1b. Anal fin spines 2–4 4

2a. Dorsal spines 12–13; side of body with irregular dark vertical bars .... *Archoplites interruptus*

Sacramento Perch, *Archoplites interruptus* (Girard, 1854). 34 mm, A08-460, Klamath River. Introduced in Klamath River and Upper Klamath Basin.

2b. Dorsal spines 5–8; side of body with spots or bars 3

3a. Dorsal fin spines 5–6; usually distinct vertical bars on side of body; dorsal fin base less than distance from back of eye to dorsal fin origin .... *Pomoxis annularis*

White Crappie, *Pomoxis annularis* Rafinesque, 1818. 135 mm, OS 18437, Willamette River. Introduced widely.

3b. Dorsal fin spines 7–8; irregular dark spots on side of body; dorsal fin base about equal to distance from back of eye to dorsal fin origin .... *Pomoxis nigromaculatus*

Black Crappie, *Pomoxis nigromaculatus* (Lesueur, 1829). Illustration by Joe Tomelleri. Introduced widely.

4a. Lateral line scale rows < 55; body deep, about 1/3 SL; side of body with uniform coloration of vertical bands 5

4b. Lateral line scale rows > 55; body elongate, depth less than 1/3 SL; side of body with dark lateral stripe..................... 9

5a. Cheek and preopercle with 3–5 dark bands radiating from eye; tongue with tooth patch .... *Lepomis gulosus*

Warmouth, *Lepomis gulosus* (Cuvier, 1829). Female, 105 mm, OS 18439, Willamette River. Introduced widely.

5b. Cheek and preopercle without dark bands radiating from eye; tongue without tooth patch... 6

6a. Pectoral fin long, pointed; when bent forward, extending beyond front of eye.. 7

6b. Pectoral fin short and rounded, when bent forward, not extending beyond front of eye; green wavy lines on cheek; leading edge of anal fin yellow; gill rakers long and slender, length of each more than twice width; opercular tab elongate and stiff to the edge; mouth large .... *Lepomis cyanellus*

Green Sunfish, *Lepomis cyanellus* Rafinesque, 1819. 95 mm, OS 18814, Willamette River. Introduced widely.

7a. Posterior dorsal fin with dark blotch near base; opercle with posterior dark tab without a light border .... *Lepomis macrochirus*

Bluegill, *Lepomis macrochirus* Rafinesque, 1819. *Above*, male, 133 mm, OS 18813; *below*, female, 113 mm, OS 18438, both Willamette River. Introduced widely.

7b. Posterior dorsal fin without a dark blotch near base; opercle with posterior tab red or orange and with light margin 8

8a. Posterior dorsal fin speckled; cheek with blue and orange stripes; gill rakers short, length of each less than twice the width; orange "seedlike" spots on side .... *Lepomis gibbosus*

Pumpkinseed, *Lepomis gibbosus* (Linnaeus, 1758). *Above*, female, 91 mm; *below*, male, 112 mm, both OS 18440, Willamette River. Introduced widely.

8b. Posterior dorsal fin without speckles; cheek without stripes ....*Lepomis microlophus*

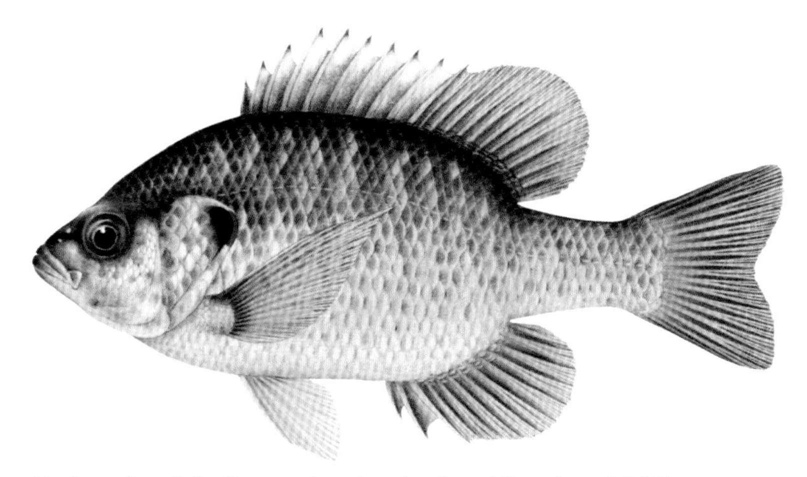

Redear Sunfish, *Lepomis microlophus* (Gunther, 1859). Illustration by Joe Tomelleri. Introduced in Willamette River.

9a. Adults with dark lateral stripe; deep notch separating spinous and soft dorsal fins; last dorsal spine less than half length of longest spine; upper jaw extending beyond posterior margin of eye; scale rows on cheek 9–12; caudal fin of juveniles with light base and dark posterior margin .... *Micropterus salmoides*

Largemouth Bass, *Micropterus salmoides* (Lacepede, 1802). 150 mm, OS 18812, Willamette River. Introduced widely.

9b. Adults with or without dark lateral stripe, and if present, containing a series of blotches; spinous and soft dorsal fins connected by membranous shallow notch; last dorsal spine more than half length of longest spine; upper jaw extending to middle or to posterior margin of eye; scale rows on cheek more than 13; caudal fin of juveniles with light yellow base, dark midregion, and clear posterior margin 10

10a. Side of body uniform or with vertical bars; no spots on side below lateral line; scale rows above lateral line 12–13; juveniles without prominent spot at base of caudal fin .... *Micropterus dolomieu*

Smallmouth Bass, *Micropterus dolomieu* Lacepede, 1802. *Above*, 38 mm, Marys River; *below*, 157 mm, Umpqua River. Introduced widely.

10b. Side with dark lateral stripe interrupted by dark blotches; horizontal rows of spots on side below lateral line; scale rows above lateral line 7–10; juveniles often with distinct spot at base of caudal fin .... *Micropterus punctulatus*

Spotted Bass, *Micropterus punctulatus* (Rafinesque, 1819). OS 17950, Lost Creek Lake. Introduced, scattered, in Willamette, Klamath, and coastal drainages.

PERCHES
family Percidae

Two exotic percids, the Yellow Perch (*Perca flavescens*) and the Walleye (*Sander vitreus*), have been introduced in Oregon. The Yellow Perch has 6–8 anal fin rays, no canine teeth, and distinct, dark vertical bars on the side. It is widespread in lakes and ponds. The Walleye has 12–13 anal fin rays, prominent canine teeth in the jaws, a narrower body, and indistinct bands. It was introduced in Washington in the 1960s and is found in the lower Willamette and Columbia Rivers.

Walleye, *Sander vitreus* (Mitchill, 1818). Illustration by Joe Tomelleri.

Yellow Perch, *Perca flavescens* (Mitchill, 1814). *Above*, 11 mm, A98-169; *below*, 165 mm, OS 18415, both Upper Klamath Lake.

GOBIES
family Gobiidae

Amur Goby, *Rhinogobius brunneus*, a native of Asia, has recently been found in the Columbia, including the Willamette River. This appears to be a complex of several species, and more work might be needed to confirm the species identification. Gobies are bottom-oriented fish like sculpins but have 2 dorsal fins and a pelvic fin modified as a sucking disk. About 40 individuals were recently collected between river miles 20 and 57 in the Willamette River by Stan Gregory and Randy Wildman.

Amur Goby, *Rhinogobius brunneus* (Temminck & Schlegel, 1845).
39 mm, OS 19258, Willamette River.

SURF PERCHES
family Embiotocidae

The surf perches are a marine group with 1 member, the Shiner Perch (*Cymatogaster aggregata*), common in estuarine and freshwater. They do not spawn in freshwater.

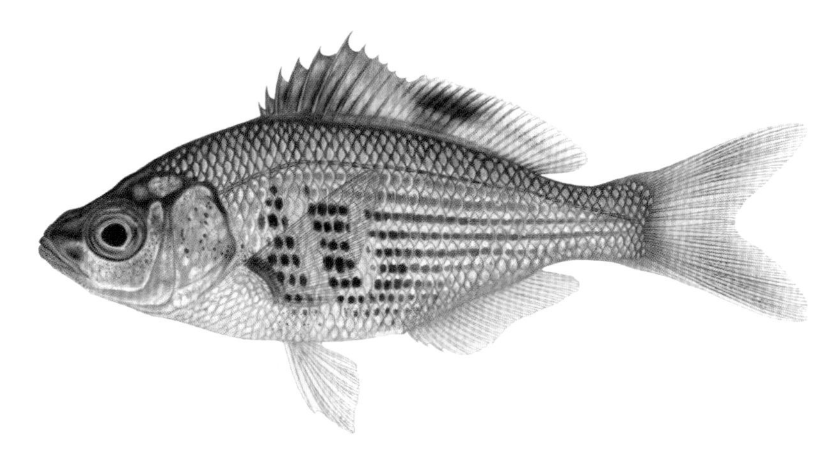

Shiner Perch, *Cymatogaster aggregata* Gibbons, 1854. Illustration by Joe Tomelleri.

FLOUNDERS
family Pleuronectidae

The flounders are a marine group with 1 member, the Starry Flounder (*Platichthys stellatus*), common in estuarine and the lower freshwater areas of coastal rivers, including the lower Willamette and lower Columbia. They do not spawn in freshwater.

Starry Flounder, *Platichthys stellatus* (Pallas, 1787).

GLOSSARY

Abdomen, abdominal—belly; referring to the belly
Adipose fin—a small fleshy fin lacking fin rays, located behind dorsal fin
Adnate—joined together
Ammocoete—blind larval stage of lampreys
Anal fin—median fin behind the vent
Anal fin origin—the most anterior point of the anal fin base
Anterior—relating to the front of a body or structure
Axillary process—an enlarged scalelike structure located at the insertion of the pectoral or pelvic fin
Barbel—an elongated fleshy projection, usually on the mouth
Base—the part of a structure, such as a fin, that is connected to the body
Basibranchial teeth—teeth located on the basibranchial, a long bone behind the tongue
Bicuspid—having 2 cusps or points
Branchial—referring to gills
Branchial aperture—gill opening
Branchiostegals, branchiostegal ray(s)—bony rays supporting the gill membranes under and behind the lower jaw
Caudal—referring to the tail; the caudal fin is the tail fin; the caudal peduncle is the region of the body between the end of the anal fin and the base of the caudal fin
Compressed—flattened laterally
Confluent—joined together
Ctenoid scale—rough scale, as found on bass
Cycloid scale—smooth scale, as found on trout
Deciduous—temporary, falling off
Depressed—flattened from top to bottom
Distal—remote from the point of attachment
Dorsal fin—the fin or fins on a fish's back

Dorsal fin origin—the most anterior point of the dorsal fin base
Elongate—extended, drawn out
Emarginate—with the margin slightly hollowed
Endemic—restricted to a particular region
Epurals—free bones lying above the caudal vertebrae
Estuarine—living in estuaries
Falcate—scythe-shaped; long, narrow, and curved
Finfold—embryonic tissue that develops into a fin
Frenum—tissue connecting snout to upper jaw such that there is no groove present between snout and upper jaw
Fusiform—tapering toward both ends
Ganoid scales—armor-like scales coated with ganoin, found in gars
Genital papilla, urogenital papilla—a small, fleshy papilla behind the anus through which urinary waste and gametes are released
Gill arches—the bony arches to which the gills are attached
Gill cover—a bony flap that covers the gills; the operculum
Gill filaments—a series of fleshy projections along the posterior edge of gill arches; site of gas exchange
Gill membranes—membranes covering the gill openings, attached to the branchiostegals
Gill opening—opening behind each operculum, leading to the gills
Gill rakers—a series of bony projections along the anterior edge of gill arches
Gills, branchiae—organs for removing oxygen from water or for breathing
Glossohyal—tongue bone
Gonopodium—modified anal fin of males of Poeciliidae; used to transfer sperm to the female during mating
Gular region—the region behind the chin and between the lower jaws
Head length—the distance from the tip of the snout (or upper lip) to the most posterior point of the opercular margin
Heterocercal—an asymmetrical caudal structure with a long upper lobe, as in sharks
Homocercal—an externally symmetrical caudal structure with approximately equal upper and lower lobes, as in most bony fishes
Hypural plate—the flattened bony plates at the posterior end of the vertebral column
Ichthyoplankton—fish eggs and larvae

Imbricate—overlapping like the shingles on a roof, in reference to scales
Inferior—referring to the lower side, usually of the head
Infraorbital—referring to the area under the eye
Insertion—the point where the last ray of a fin attaches to the body
Interorbital—the space between the orbits
Isthmus—the fleshy area of the body separating the gill openings and location of the heart
Jugular—referring to the throat
Keel—a raised ridge, often on scales
Lateral—referring to the side
Lateral line—a sensory structure usually located in pored scales in a line along the side of the body
Longitudinal series (scales)—the number of scale rows above the lateral line from the first pored lateral line scale to the caudal fin base
Lunate—shaped like a crescent moon, with long upper and lower lobes
Macrophthalmia—the transformer stage of anadromous lampreys with a proportionally larger eye
Maxilla, maxillary—the upper jaw, or referring to it
Median, medially—referring to the middle
Median fins—fins located on the median line of a fish; the dorsal, anal, and caudal fins
Melanophore—a cell (chromatophore) containing melanin or other black pigment
Morphology—form and structure of an organism
Myomeres—blocks of muscle forming a sideways *W* and corresponding to the number of vertebrae, easily seen in larval fishes and fish fillets
Nape—upper body behind the head and in front of the dorsal fin
Nasal—referring to the nostrils
Notochord—a rudimentary embryonic spinal column
Nuchal—referring to the nape
Oblique—referring to a sharp angle, as in an oblique mouth
Occipital—referring to the posterior part of the skull
Opercle—the large bone that forms the upper posterior part of the operculum
Opercular spine—spine(s) projecting from the operculum
Operculum—series of bones that covers the gills

Orbit—the eye socket

Origin—the point where the first ray of a fin attaches to the body; the most anterior point of a fin base

Palate—the roof of the mouth

Palatines—the bones on each side of the palate

Papilla (papillae)—small fleshy projection(s)

Parr marks—a series of round to oval dark marks on the sides of juvenile salmon and trout

Pectoral fins—the anterior or uppermost of the paired fins

Pelvic fins—paired fins behind or below the pectoral fins

Peritoneum—lining of the body cavity

Pharyngeal bones—last pair of gill arches, without filaments

Pharyngeal teeth—often present on pharyngeal bones, creating "jaws" in the esophagus or gullet

Pharynx—posterior part of throat, into which the gill slits open

Posterior—toward the hind end of the fish

Premaxillaries—2 bones forming all or part of the upper jaw

Preopercle, preoperculum—part of the operculum; bone between the cheek and the opercle

Preopercular spine—spine projecting from the preopercle

Procurrent—small fin rays in front of primary caudal fin rays

Protrusible—capable of extending forward, often referring to the jaws of fishes

Proximal—nearest

Pterygiophore—a series of internal cartilage or bone structures supporting a median fin ray or spine

Pyloric caeca—fingerlike pouches connected with the gut

Ray—a paired, segmented, often branched rod that supports a fin

Rostrum—a projecting snout or beak

Scute—any external horny or bony plate or scalelike structure, usually with a sharp raised middle ridge

Serrate—notched like a saw

Soft dorsal—the posterior part of the dorsal fin that is composed of rays

Spine—an unsegmented, unbranched, sharp rod that supports the anterior portions of some dorsal and anal fins

Spinous dorsal—the anterior part of the dorsal fin supported by spines

Standard length (SL)—the length of a fish as measured from the tip of the snout to the posterior edge of the caudal fin plate (hypurals)

Suborbital—below the eye
Superior—above or on the upper surface
Supraorbital—above the eye
Swimbladder—a sac filled with gas, lying beneath the backbone
Symphysis—point of junction of 2 sides of the jaw
Teleost—a member of Teleostei, an infraclass containing most of the bony fishes
Terminal—at the end
Thoracic—referring to the chest; subthoracic is just below (behind) the chest
Total length (TL)—the length from the tip of the snout to the posterior tip of the caudal fin
Truncate—squared off
Tubercles—small, deciduous, pointed structures on scales and fins that usually develop during breeding season in some fishes
Type locality—the location from which a type specimen was collected
Type specimen—also called holotype, the specimen that is the unequivocal bearer of the scientific name
Vent—the external opening of the alimentary canal; the anus
Ventral or **ventrum**—referring to the abdominal or lower surface
Ventral fins—pelvic fins; paired fins behind or below the pectoral fins
Vertical fins—fins on the median line of the body; the dorsal, anal, and caudal fins
Vomer—a bone forming the front part of the roof of the mouth, often with teeth
Weberian apparatus—a complex structure in the Ostariophysi that includes modified anterior vertebrae used to improve hearing

REGIONAL FISH REFERENCES

Behnke, R. J. 2002. *Trout and Salmon of North America.* New York: Free Press. 359 pp.

Bond, C. E. 1994. *Keys to Oregon Freshwater Fishes.* Rev. ed. Corvallis: Oregon State University Book Stores. 51 pp.

Boschung, H. T., and R. L. Mayden. 2004. *Fishes of Alabama.* Washington, DC: Smithsonian Books. 736 pp.

Cavender, T. M. 1986. "Review of Fossil History of North American Freshwater Fishes." In *The Zoogeography of North American Freshwater Fishes*, edited by C. H. Hocutt and E. O. Wiley, 699–724. New York: John Wiley and Sons. 866 pp.

Dauble, D. D. 2009. *Fishes of the Columbia Basin: A Guide to Their Natural History and Identification.* Sandpoint, ID: Keokee Books. 210 pp.

Lampman, B. H. 1946. *The Coming of the Pond Fishes: An Account of the Introduction of Certain Spiny-Rayed Fishes and Other Exotic Species, into the Waters of the Lower Columbia River Region and the Pacific Coast States.* Portland, OR: Binfords and Mort. 177 pp.

McPhail, J. D. 2007. *The Freshwater Fishes of British Columbia.* Edmonton: University of Alberta Press. 620 pp.

Moyle, P. B. 2002. *Inland Fishes of California.* Rev. ed. Berkeley: University of California Press. 502 pp.

Page, L. M., and B. M. Burr. 2011. *Peterson Field Guide to Freshwater Fishes.* 2nd ed. Boston: Houghton-Mifflin. 663 pp.

Page, L. M., H. Espinosa-Pérez, L. T. Findley, C. R. Gilbert, R. N. Lea, N. E. Mandrak, R. L. Mayden, and J. S. Nelson. 2013. *Common and Scientific Names of Fishes from the United States, Canada, and Mexico.* 7th ed. American Fisheries Society, Special Publication 34. 243 pp.

Pollard, W. R., G. F. Hartman, C. Groot, and P. Edgell. 1997. *Field Identification of Coastal Juvenile Salmonids.* Madeira Park, BC: Harbour Publishing. 32 pp.

Simpson, J., and R. Wallace. 1978. *Fishes of Idaho*. Moscow: University of Idaho Press. 237 pp.

Smith, G. R., T. E. Dowling, K. W. Gobalet, T. Lugaski, D. K. Shiozawa, and R. P. Evans. 2002. "Biogeography and Timing of Evolutionary Events among Great Basin Fishes." *Smithsonian Contributions to the Earth Sciences* 33:175–234.

Wallace, R. L., and D. W. Zaroban. 2013. *Native Fishes of Idaho*. Bethesda, MD: American Fisheries Society. 216 pp.

Warren, M. L., and B. M. Burr, eds. 2014. *Freshwater Fishes of North America*. Vol. 1, *Petromyzontidae to Catostomidae*. Baltimore: Johns Hopkins University Press. 644 pp.

Williams, J. E., G. R. Giannico, and B. Winthrow-Robinson. 2014. *Field Guide to Common Fish of the Willamette Valley Floodplain*. Oregon State University Extension Service, EM9091. 41 pp.

Wydoski, R. S., and R. R. Whitney. 2003. *Inland Fishes of Washington*. Rev. ed. Bethesda, MD: American Fisheries Society. 322 pp.

REGIONAL DRAINAGE/RIVER REFERENCES

Aalto, K. R., W. D. Sharp, and P. R. Renne. 1998. "^{40}Ar/^{39}Ar Dating of Detrital Micas from Oligocene-Pleistocene Sandstones of the Olympic Peninsula, Klamath Mountains, and Northern California Coast Ranges: Provenance and Paleodrainage Patterns." *Canadian Journal of Earth Sciences* 35:735–45. doi:10.1139/e98-025.

Benke, A. C., and C. E. Cushing. 2005. *Rivers of North America*. Burlington, MA: Elsevier Academic Press. 1168 pp.

O'Connor, J. E., and V. R. Baker. 1992. "Magnitudes and Implications of Peak Discharges from Glacial Lake Missoula." *Geological Society of America Bulletin* 104:267–79.

Palmer, T. 2014. *Field Guide to Oregon Rivers*. Corvallis: Oregon State University Press. 320 pp.

Smith, G. R. 1981. "Late Cenozoic Freshwater Fishes of North America." *Annual Review of Ecology and Systematics* 12:163–93.

Smith, G. R., N. Morgan, and E. Gustafson. 2000. "Fishes of the Mio-Pliocene Ringold Formation, Washington: Pliocene Capture of the Snake River by the Columbia River." *University of Michigan Papers on Paleontology* 32:1–47.

Taylor, D. W. 1985. "Evolution of Freshwater Drainages and Molluscs in Western North America." In *Late Cenozoic History of the Pacific Northwest*, edited by C. J. Smiley, 265–321. San Francisco: AAAS. 417 pp.

INDEX TO COMMON AND SCIENTIFIC NAMES